METHODS IN MOLECULAR BIOLOGY™

Series Editor
John M. Walker
School of Life Sciences
University of Hertfordshire
Hatfield, Hertfordshire, AL10 9AB, UK

For further volumes:
http://www.springer.com/series/7651

Immune Homeostasis

Methods and Protocols

Edited by

Andrew L. Snow

Department of Pharmacology, Uniformed Services, University of the Health Sciences, Bethesda, MD, USA

Michael J. Lenardo

Laboratory of Immunology, National Institutes of Health, NIAID, Bethesda, MD, USA

Humana Press

Editors
Andrew L. Snow
Department of Pharmacology
Uniformed Services
University of the Health Sciences
Bethesda, MD, USA

Michael J. Lenardo
Laboratory of Immunology
National Institutes of Health
NIAID
Bethesda, MD, USA

ISSN 1064-3745 ISSN 1940-6029 (electronic)
ISBN 978-1-4939-5968-6 ISBN 978-1-62703-290-2 (eBook)
DOI 10.1007/978-1-62703-290-2
Springer New York Heidelberg Dordrecht London

Printed on acid-free paper

Humana Press is a brand of Springer
Springer is part of Springer Science+Business Media (www.springer.com)

Preface

Immunology constitutes the study of an extremely powerful and dynamic system for sensing, containing, and eliminating foreign pathogens with minimal damage to the host. The success of this process is contingent on the activation and rapid clonal expansion of lymphocytes and accessory cells that participate in executing an effective, pathogen-tailored adaptive immune response. Moreover, the proper regulation and contraction of these potent immune cells during and after the immune response is just as important to the health of the host. Many questions remain unresolved as we continue to characterize and define the nature of normal immune homeostasis and determine how these processes are dysregulated in immunodeficiency, as well as in autoimmune and lymphoproliferative disorders.

This volume of Methods in Molecular Biology on "Immune Homeostasis" focuses on experimental techniques for measuring and analyzing immune cell dynamics, with a particular emphasis on examining lymphocyte programmed cell death in different contexts. Chapters 1–7 of the book aim to equip the researcher with detailed protocols for studying various pathways of apoptosis and necrosis in different types of hematopoietic cells, both in vitro and in vivo. Chapters 8–12 offer several methods for studying the maintenance of lymphocyte populations in the steady state or following infectious challenges in both mice and humans. Chapters 13–14 offer technical insights into state-of-the-art genomics tools for investigating gene sequence and expression variation, both in small subsets of lymphocytes and on a genome-wide scale, in individuals challenged with infections or inherited immune disorders. Chapter 15 provides an example of a flow cytometric assay employed for the diagnosis of specific immunodeficiencies. Finally, Chapter 16 outlines an assay for investigating regulatory T cell-mediated suppression of immune responses.

Although no single volume can encompass the full range of methods available for studying immune responses, this volume provides a valuable "starter" toolkit for basic and clinical scientists interested in examining various aspects of immune homeostasis in both normal and disease-related contexts. We hope the clever researcher will envision how the techniques outlined herein may be directly applied, altered, or combined to address his/her particular question of interest. With these techniques in hand, we wish you continued success in studying the dynamic processes that contribute to homeostasis of the immune system.

Bethesda, MD, USA — *Andrew L. Snow*
Bethesda, MD, USA — *Michael J. Lenardo*

Contents

Contributors

R. ANTHONY BARNITZ • *Department of Pediatric Oncology, Dana–Farber Cancer Institute, Harvard Medical School, Boston, MA, USA*
MAGDALENA A. BERKOWSKA • *Department of Immunology, Erasmus MC, University Medical Center, Rotterdam, The Netherlands*
NICOLAS BIDÈRE • *INSERM UMR_S 1014, Hôpital Paul Brousse, Villejuif, France; Université Paris-Sud P11, Orsay, France*
JACK J. BLEESING • *Division of Bone Marrow Transplantation and Immune Deficiency, Cancer and Blood Diseases Institute, Cincinnati Children's Hospital Medical Center, Cincinnati, OH, USA*
JOSÉ A.M. BORGHANS • *Department of Immunology, University Medical Center Utrecht, Utrecht, The Netherlands*
MIRJAM VAN DER BURG • *Department of Immunology, Erasmus MC, University Medical Center, Rotterdam, The Netherlands*
FRANCIS KA-MING CHAN • *Immunology and Virology Program, Department of Pathology, Center for AIDS Research, University of Massachusetts Medical School, Worcester, MA, USA; Diabetes and Endocrinology Research Center, University of Massachusetts Medical School, Worcester, MA, USA*
MIN CHEN • *Department of Pathology and Immunology, Baylor College of Medicine, Houston, TX, USA*
NELIA CORDEIRO • *INSERM UMR_S 1014, Hôpital Paul Brousse, Villejuif, France; Université Paris-Sud P11, Orsay, France*
JACQUES J.M. VAN DONGEN • *Department of Immunology, Erasmus MC, University Medical Center, Rotterdam, The Netherlands*
ALEXANDRA H. FILIPOVICH • *Division of Bone Marrow Transplantation and Immune Deficiency, Cancer and Blood Diseases Institute, Cincinnati Children's Hospital Medical Center, Cincinnati, OH, USA*
KWESI FRIMPONG-BOATENG • *Doctoral Program in Chemical and Biological Sciences, Kellogg School of Science and Technology, The Scripps Research Institute, La Jolla, CA, USA; Department of Immunology and Microbial Science, The Scripps Research Institute, La Jolla, CA, USA*
W. NICHOLAS HAINING • *Department of Pediatric Oncology, Dana–Farber Cancer Institute, Harvard Medical School, Boston, MA, USA; Division of Hematology/Oncology, Children's Hospital, Harvard Medical School, Boston, MA, USA*
DAVID HILDEMAN • *Division of Cellular and Molecular Immunology, Department of Pediatrics, University of Cincinnati, Cincinnati, OH, USA; Cincinnati Children's Hospital Medical Center, Cincinnati, OH, USA*
LILY HUANG • *Department of Pathology and Immunology, Baylor College of Medicine, Houston, TX, USA*

SABRINA IMAM • *Department of Pediatric Oncology, Dana–Farber Cancer Institute, Harvard Medical School, Boston, MA, USA*
GIL KATZ • *Department of Pharmacology, Uniformed Services University of the Health Sciences, Bethesda, MD, USA*
SEMA KURTULUS • *Division of Immunobiology, Department of Pediatrics, University of Cincinnati, Cincinnati, OH, USA; Cincinnati Children's Hospital Medical Center, Cincinnati, OH, USA*
BERNICE LO • *Laboratory of Immunology, National Institute of Allergy and Infectious Diseases, National Institutes of Health, Bethesda, MD, USA*
HEATHER E. LYNCH • *Duke Human Vaccine Institute, Duke University Medical Center, Durham, NC, USA; Department of Medicine, Duke University Medical Center, Durham, NC, USA*
DEREK C. MACALLAN • *Division of Clinical Sciences, Infection & Immunity Research Centre, St George's, University of London, London, UK*
REBECCA A. MARSH • *Division of Bone Marrow Transplantation and Immune Deficiency, Cancer and Blood Diseases Institute, Cincinnati Children's Hospital Medical Center, Cincinnati, OH, USA*
CHRISTOPHER E. MARTIN • *Doctoral Program in Chemical and Biological Sciences, Kellogg School of Science and Technology, The Scripps Research Institute, La Jolla, CA, USA; Department of Immunology and Microbial Science, The Scripps Research Institute, La Jolla, CA, USA*
KENTA MORIWAKI • *Immunology and Virology Program, Department of Pathology, Center for AIDS Research, University of Massachusetts Medical School, Worcester, MA, USA; The Diabetes and Endocrinology Research Center, University of Massachusetts Medical School, Worcester, MA, USA*
JOAO BOSCO OLIVEIRA • *Pediatrics Department, Medical Institute Prof. Fernando Figueira-IMIP Recife, PE Brazil*
MADHU RAMASWAMY • *Immunoregulation Unit, Autoimmunity Branch, National Institute of Arthritis and Musculoskeletal and Skin Diseases, National Institutes of Health, Bethesda, MD, USA*
MARÍA JOSÉ DE ROSA • *Immunology and Virology Program, Department of Pathology, Center for AIDS Research, University of Massachusetts Medical School, Worcester, MA, USA; The Diabetes and Endocrinology Research Center, University of Massachusetts Medical School, Worcester, MA, USA*
GREGORY D. SEMPOWSKI • *Duke Human Vaccine Institute, Duke University Medical Center, Durham, NC, USA; Department of Medicine, Duke University Medical Center, Durham, NC, USA*
ANDREW L. SNOW • *Department of Pharmacology, Uniformed Services University of the Health Sciences, Bethesda, MD, USA*
DARINA S. SPASOVA • *Doctoral Program in Chemical and Biological Sciences, Kellogg School of Science and Technology, The Scripps Research Institute, La Jolla, CA, USA; Department of Immunology and Microbial Science, The Scripps Research Institute, La Jolla, CA, USA*
JOHN C. STONE • *Department of Immunology and Microbial Science, The Scripps Research Institute, La Jolla, CA, USA*
HELEN C. SU • *Human Immunological Diseases Unit, Laboratory of Host Defenses, National Institute of Allergy and Infectious Diseases, National Institutes of Health, Bethesda, MD, USA*

Charles D. Surh • *Academy of Immunology and Microbiology (AIM), Institute of Basic Science (IBS) Pohang University of Science and Technology (POSTECH), Pohang, Republic of Korea*
Kiki Tesselaar • *Department of Immunology, University Medical Center Utrecht, Utrecht, The Netherlands*
Dat Q. Tran • *Division of Pediatric Research Center, Department of Pediatrics, The University of Texas Medical School at Houston, Houston, TX, USA*
Jin Wang • *Department of Pathology and Immunology, Baylor College of Medicine, Houston, TX, USA*
Liset Westera • *Department of Immunology, University Medical Center Utrecht, Utrecht, The Netherlands*
Kathleen Yates • *Department of Pediatric Oncology, Dana-Farber Cancer Institute, Harvard Medical School, Boston, MA, USA*
Menno C. van Zelm • *Department of Immunology, Erasmus MC, University Medical Center, Rotterdam, The Netherlands*
Yan Zhang • *Division of Clinical Sciences, Infection & Immunity Research Centre, St George's, University of London, London, UK*
Yu Zhang • *Human Immunological Diseases Unit, Laboratory of Host Defenses, National Institute of Allergy and Infectious Diseases, National Institutes of Health, Bethesda, MD, USA*
Lixin Zheng • *Laboratory of Immunology, National Institute of Allergy and Infectious Diseases, National Institutes of Health, Bethesda, MD, USA*

Chapter 1

Quantitating Lymphocyte Programmed Cell Death In Vitro Using Simple Kill Assays

Lixin Zheng

Abstract

Programmed cell death is essential to maintaining lymphocyte homeostasis during the contraction phase of the immune response. Activated lymphocytes become susceptible to a variety of programmed cell death (PCD) stimuli over the course of a typical immune response. This chapter outlines two simple approaches for measuring programmed cell death of lymphocytes cultured in vitro, regardless of the stimulus provided. These techniques exploit changes in plasma membrane integrity and/or mitochondrial membrane potential that are characteristic of cells undergoing PCD. The detection methods discussed are generally applicable for assessing cell death in several contexts, expanded upon in further detail in subsequent chapters.

Key words: Cell viability, apoptosis, TCR, RICD (re-stimulation induced cell death), CWPD (cytokine-withdrawal induced programmed cell death), flowcytometry

1. Introduction

A typical adaptive immune response includes the activation and proliferation of antigen-reactive lymphocytes at beginning, exercise of effector functions in the middle, and contraction of the "mission-accomplished" effector lymphocytes and other immune cells at the end. It is evident that homeostasis of the immune system relies on normal programmed cell death (PCD) at the contraction phase of an immune response. Abnormal PCD in lymphocytes and other immune cells can cause autoimmune lympho-proliferative syndrome and related autoimmune diseases (1). Therefore, assessment of PCD is essential for studying the mechanisms underlying the homeostatic regulation of the immune system.

Andrew L. Snow and Michael J. Lenardo (eds.), *Immune Homeostasis: Methods and Protocols*, Methods in Molecular Biology, vol. 979, DOI 10.1007/978-1-62703-290-2_1, © Springer Science+Business Media New York 2013

Kill assays, also called death assays, are defined to assess the number of cells that die over a period of time in experimental samples. As an opposite indicator of cellular viability, measuring cell death has been widely used for analyzing activation-induced cell death (AICD), T-cell receptor (TCR) restimulation-induced cell death (RICD), cytokine-withdrawal-induced cell death (CWID), and other PCD mechanisms in lymphocytes. For practical reasons, most kill assays are conducted in vitro. The resulting data can be presented in different ways, such as the percentage of death in total cell counts, the absolute count of live or dead cells, and alternatively, the percentage of cell loss.

Over recent years, there are many cell-death/viability assay kits available on the market. These methods can largely be divided into two categories consistent with their mechanism of action. One is based on measurement of cell membrane integrity, including the use of dyes like trypan blue (2), propidium iodide (PI) (3), and 7-aminoactinomycin D (7-AAD) (4) that are membrane impermeable for healthy cells; positive staining of cells by these dyes therefore marks cell death. The other category covers a variety of mechanisms involving the use of membrane-permeable and -impermeable molecules or dyes. To give a few examples, annexin V stains extracellular exposed phosphatidylserine on the plasma membrane of early apoptotic cells (5); fluorescent caspase substrates indicate caspase activation (6); or metabolic dyes stain live and dead cells with different colors (7). These measurements exploit unique features of dead or dying cells aside from loss of membrane integrity.

Alternatively, the terminal transferase-mediated dUTP nick end-labeling (TUNEL) assay can be used to identify apoptotic cells in tissues (8). Nevertheless, the development of applicable in vivo kill assays has still been a challenge for the field. With the increasing availability of new cell death detection reagents and sophisticated multiphoton microscopy techniques, we can expect to see the emergence of new assay systems in the near future that can detect lymphocyte programmed death in vivo in animal models, or in ex vivo tissue sections from patient samples.

This chapter focuses on two easily applicable and very reliable approaches for assessing cell death in vitro. The first is based on staining samples with membrane-impermeable dyes and using microscopes or microscope-derived cell counting instruments to distinguish live and dead cells. The other approach uses flow cytometry—for measuring viability and death by staining with particular dyes. These two basic techniques should be adequate for in vitro studies of RICD, CWID, and other cell-death/viability analyses of the immune system. Here we outline the basic premise of each approach and discuss some considerations for appropriately using these methods.

2. Membrane-Dye Exclusion Assay

The trypan blue exclusion assay, which was introduced 87 years ago for distinction of live and dead cells in vitro (2), is still one of the simplest, most reliable, and widely used methods of its kind. This protocol needs nothing more than a bottle of 0.04% trypan blue dye in PBS, a hemacytometer, and an ordinary light microscope. The basic protocol provides that a10 μl single-cell suspension, like cultured lymphocytes or a well-prepared single-cell suspension of adherent cells, is mixed with 90 μl of 0.04% trypan blue in PBS. Then a 10 μl of the mixture is loaded onto a hemacytometer beneath a cover slide, and is then subjected to light microscopic inspection. Live cells having an intact membrane will exclude trypan blue from staining, and therefore will appear shiny and bright, with their shapes variable depending on the natural characteristics of the cells. In contrast, dead or dying cells that lose membrane integrity become membrane permeable for the dye and stain blue. Thus, live and dead cells are easily distinguished under a microscope. As a common practice, live and/or dead cells are counted from the 4 large corner squares, each containing 16 smaller squares in a hemacytometer, by including all cells within the area as well as those across the top and left border lines but not the right and bottom lines (Fig. 1). The total count is divided by 4, giving the number of live or dead cells per 1/10 of a cubic millimeter. Therefore, this number ($\times 10^5$) yields the concentration of live/dead cells per milliliter (ml). By following these simple steps, one can collect absolute numbers as well as percentage of live and dead cells in a given sample: Cell counts (per ml) = total counts of the 4 large corners/4 $\times 10^4 \times$ dilution factor.

Another advantage of using the light microscope to inspect sample cells is to observe morphological changes characteristic of dying cells. For example, before losing membrane integrity, apoptotic cells may still be capable of excluding trypan blue stains; however, these dying cells may manifest membrane blebbing, cytoplasm/nuclear condensation, and segregation under the light microscope. In many occasions, a careful microscopic inspection of the dying cells can lead to a definitive conclusion for what form of cell death is occurring in the samples under examination.

Alternatively, many other dyes such as Giemsa stain or assorted fluorescent dyes have also been used for this purpose and can provide sensitive and accurate measurements by using conventional and fluorescent microscopic procedures, such as for detecting apoptotic versus necroptotic cell death. One way of achieving this is to stain the sample cells with a combination of two DNA-binding fluorescent dyes, one membrane permeable such as acridine orange, Hoechst 33342, and DAPI and the other impermeable like ethidium bromide (EB), 7-AAD, and PI. Using a simple fluorescent

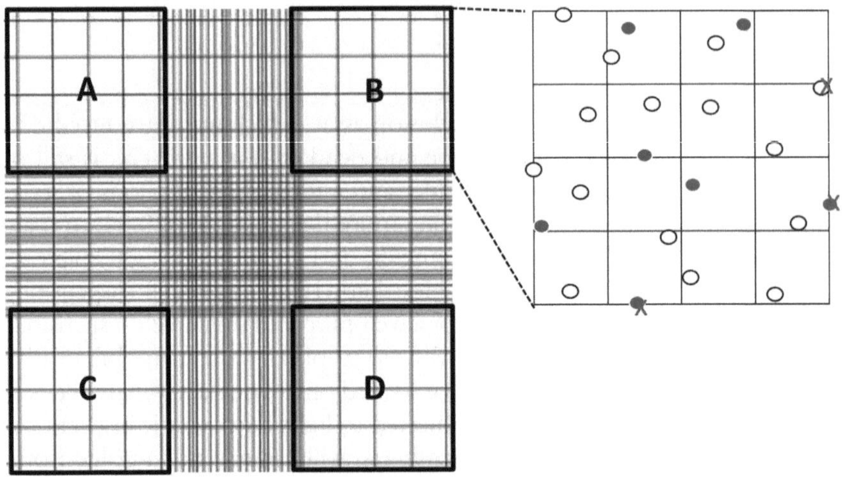

Fig. 1. Area view of a hemacytometer for counting lymphocytes and other cells. The cell concentration is calculated by counting all cells in 4 large corner boxes (**a**, **b**, **c**, and **d**, including the *top* and *left*, excluding the bottom and right line-crossing cells). The obtained number/4 × 10^4 = cells per ml.

microscope, one may be able to distinguish early apoptosis from late apoptosis and necrosis through observation of nuclear morphology changes in cells that have undamaged or broken plasma membrane integrity. In principle, one can also obtain absolute numbers of live, apoptotic, and dead cells by following the same procedure outlined for trypan blue exclusion.

Automatic cell counting machines (e.g., Countess, Cellometer) are now available on the market, allowing investigators to carry out cell-viability analysis similar to the trypan blue exclusion assay, with additional options of using elaborated protocols, including a combination of permeable and impermeable dyes of choice. By setting the machine with defined parameters, such as the size and fluorescence of cells, investigators can automatically collect images from the test samples and then obtain quantified counts of viable and dead cells. The downside of these machines is that they are expensive compared to a conventional microscope, and require continuous supplies of the nonreusable sample slides.

3. Flow Cytometry-Based Cell Death Assays

3.1. Basic PI Exclusion Assay

Taking advantage of a flow cytometer with a stable sample flow rate that can simultaneously measure multiple perimeters, a flow cytometry-based kill assay was developed by the Lenardo lab in the early 1990s that has been successfully used to examine the viability or death of single-cell suspension samples (3). This method is especially apt for analyzing lymphocytes and other leukocytes. In principle, it should be easily adaptable for most other cultured cells, given that the cultured cells are appropriately prepared, such as using trypsin/EDTA mix to shift adherent cells into a single-cell suspension. This kill assay can be conducted either manually by putting samples one by one onto a flow cytometer or through an automatic sampler attached with a machine. The basic idea is to count viable and dead cells by utilizing a membrane-impermeable dye such as PI or 7-AAD. Since PI or 7-AAD are excluded from healthy cells that have intact plasma membranes, one can simply set up a flow cytometer with forward scatter (FSC) in linear range to measure cell size versus an FL2-H or FL3-H channel in log range to measure PI/7-AAD staining. The crux of the assay is to set up the acquisition of the machine to collect events for a constant period of time, e.g., 30 s per sample, so that both absolute PI-negative (viable cell) counts and PI-positive (dead or dying cell) events per unit of time can be acquired (see Fig. 2a). Subsequent chapters utilize this basic principle to assess sensitivity to specific apoptotic stimuli.

For a typical kill assay, a standard concentration of cells (e.g., 2.5×10^5/ml) is seeded into a multi-well plate. The cells are then subjected to different doses of an apoptotic stimulus (e.g., anti-FAS stimulation, see Chapter 4) for a defined period of time in a cell-culture CO_2 incubator. For suspension cells, the culture is pipetted up and down for a few times to retrieve all of the cells in suspension before sample collection. Then a minimum of 200 μl of constant volume is transferred into an FACS tube containing 10 μl of the PI stock solution (typically 1–100 μg/ml). In this case, no extra washing step is needed. For adherent cell cultures, an appropriate detachment procedure is required to make a single-cell suspension before collecting samples for flow cytometry analysis. For example, a 0.05 % trypsin–EDTA mix is usually applied to wells after aspirating the culture medium and washing with PBS. Washing steps should be minimized to avoid introducing artificial errors. A constant volume of FC buffer containing a final concentration of PI is added to make a final single-cell suspension for flow cytometry analysis. In both cases, a constant volume across all samples at this point is critical for generating the best interpretable data.

The prepared samples are applied for acquisition on a flow cytometer, for which the acquisition is set to collect a constant amount of time (e.g., a period of 30 s is usually sufficient). Most cytometers give the option of collecting on constant time or events; by setting the machine to stop acquisition after collecting an arbitrarily high number of events (e.g., 2 million), it allows the flow cytometer to fully collect for equal amount of time for every sample. The specific death in a given experiment is calculated by comparing the absolute number of live cells (FCS high and PI negative) in the treatment groups with that in the controls (untreated) by the following equation:

$$\text{Specific death (\%)} = (1 - \text{live counts of treated sample} / \text{live counts of control samples}) \times 100\%$$

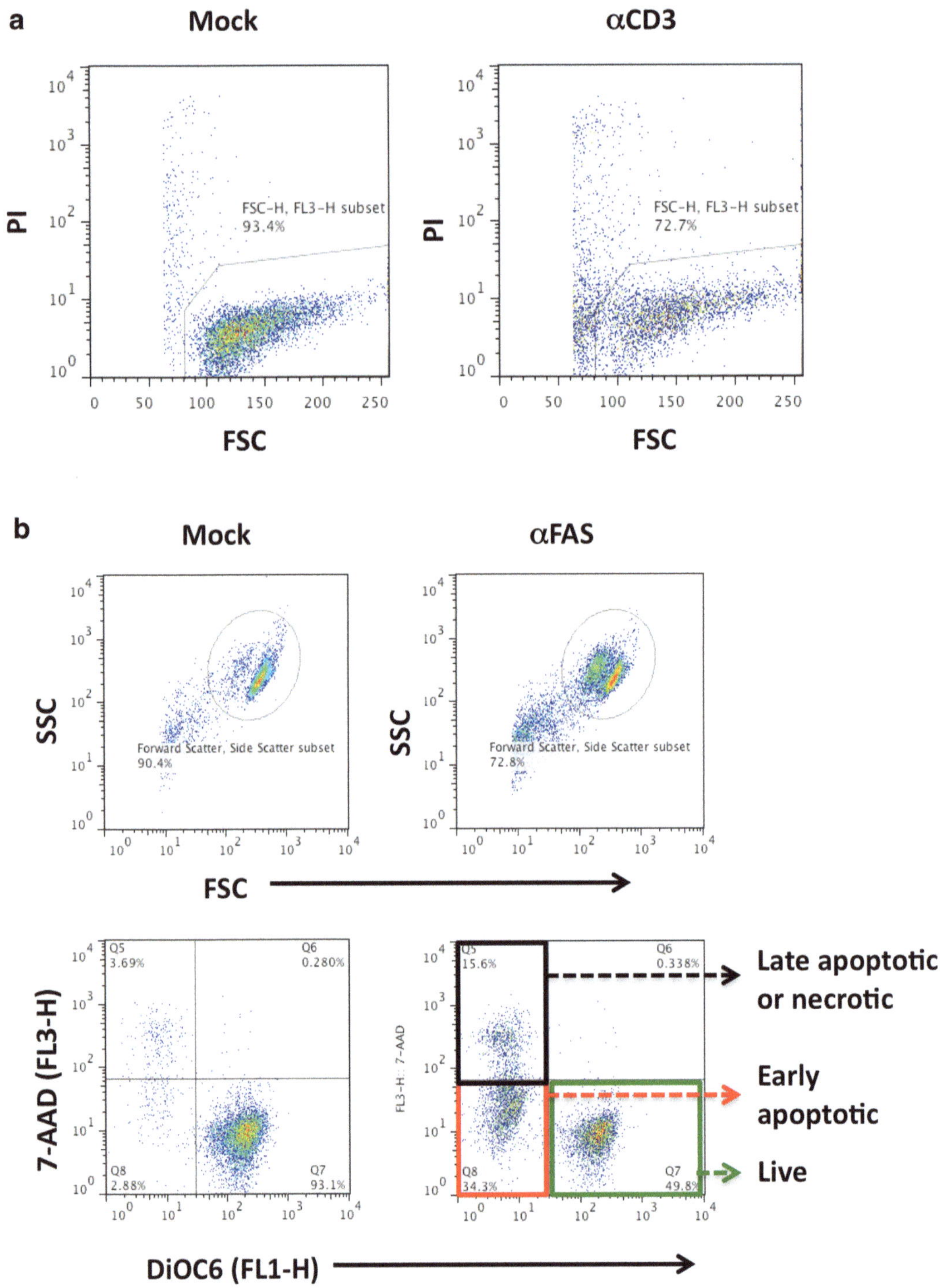

Fig. 2. FACS plots for a typical flow-based kill assay. (**a**) An example of a PI exclusion assay. A constant time (30 s) collection is implemented on the flow cytometer, plotting linear FSC versus log FL3-H (PI) for detection of the PI^- live and the PI^+ dead cells (IL-2-cultured activated PBMC). Note that the FSC/PI intermediate population represents dying cells. (**b**) Using DiOC$_6$ and 7-AAD for a kill assay. IL-2-cultured, activated human T cells were subjected to 10 ng/ml of anti-FAS stimulation for 24 h. The DiOC$_6$- and 7-AAD-stained cells were analyzed by flow cytometry. The mock-treated (*left columns*) and anti-FAS (*right columns*) cells were gated on FSC/SSC as shown at the top panels, and plotted with DiOC$_6$ versus 7-AAD for the bottom panels. Live (*green*), early apoptotic dying (*red*), and late apoptotic/necrotic dead (*black*) cells are indicated.

Table 1
Fluorescence spectra of common dyes used in kill assays

Molecules	Peak emission (nm) (range)	Notes
PI[a]	620 (~560–720)	Stains DNA in cells with damaged membrane integrity
7-AAD[b]	650 (~590–720)	Similar to PI, with lower emission spectrum
$DiOC_6$[c]	506 (~490–590)	Fluorescent intensity decreases as a function of loss of mitochondrial membrane potential ($\Delta\Psi$)
JC-1[d]	525 (~510–550) 595 (~580–610)	Fluorescence changes from red to green in response to loss of mitochondrial membrane potential ($\Delta\Psi$)
Annexin V (-FITC, PE)	518 (~495–590) 575 (~555–640)	Stains phosphatidylserine exposed on the plasma membrane of early apoptotic cells

[a] PI propidium iodide
[b] 7-AAD 7-aminoactinomycin D
[c] DiOC6 3,3′-dihexiloxocarbocyanine iodide
[d] JC-1 5,5′,6,6′-tetrachloro-1,1′,3,3′-tetraethylbenzimidazolcarbocyanine iodide

Percent specific death is often described as % cell loss, reflecting the fact of losing viable cells by experimental treatments as compared to the nontreatment controls.

3.2. Alternative Protocols of Flow Cytometry-Based Death Assays

As an extension of the basic principles in the aforementioned viability/death assay using PI, there are many derivative protocols using more than one dye to stain the cells of interest, through which investigators can detect apoptosis, necroptosis, and other possible forms of cell death. It is even possible to examine the viability of particular populations of cells by using multicolor surface staining techniques. Among the available cell death assessment dyes, PI, 7-AAD, annexin V, JC-1, and $DiOC_6$ are commonly used. Table 1 presents the parameters of these molecules regarding their potential usage for flow cytometry. PI and 7-AAD, as described before, are two membrane-impermeable dyes that are excluded by healthy cells; these dyes can intercalate and stain intracellular DNA/RNA only when cells lose plasma membrane integrity. However, the plasma membrane remains intact in the early stages of apoptosis, making it impossible to detect early apoptotic cells using these membrane-impermeable dyes.

In a standard flow cytometer (e.g., BD FACSCalibur), an argon 488 nm laser excites 7-AAD with a roughly ~30 nm narrower spectrum of emission wavelength than that of PI. This allows investigators to easily use 3 channels for measuring cell death. Annexin V is a protein that specifically binds to phosphatidylserine expressed on cell surface, a typical feature of early apoptotic cells (5, 9). Fluorochrome-conjugated annexin V reagents are commercially available from a variety of sources. $DiOC_6$ and JC-1 are two membrane-permeable fluorochromes that are retained within healthy mitochondria, accompanied by a normal mitochondrial membrane potential. Therefore, $DiOC_6$ stains healthy normal cells with high intensity in the FL-1 channel, and the intensity decreases when cells are undergoing apoptosis or responding to other cell stress signals that cause mitochondrial potential ($\Delta\Psi$) to drop. JC-1 has two different emission profiles when excited. JC-1 accumulates in normal mitochondria, forming aggregates that emit red fluorescence in the FL-2 channel. Apoptotic cells, however, featuring loss of $\Delta\Psi$, prevent JC-1 from aggregating in mitochondria, resulting in cytoplasmic distribution of the green fluorescent JC-1 monomer. It has been suggested that JC-1 staining for apoptotic cells appears to be more reliable (5, 9). However, in order to measure the change of emission wavelength during apoptosis, it will occupy 2 (FL1 and FL2) channels of a flow cytometer, limiting the number of available channels for other perimeters. $DiOC_6$ only uses the FL1 channel and in our hands, it can efficiently mark the drop of $\Delta\Psi$, allowing investigators to save a detection channel for assessment of another parameter in an experiment. Also noteworthy, $DiOC_6$ can be added into cell cultures for more than 3 days for measuring $\Delta\Psi$, with no toxicity detected (Zheng and Lenardo, unpublished data). Taken together, $DiOC_6$ appears to be a wonderful dye for measuring mitochondrial $\Delta\Psi$.

Detailed protocols utilizing these methods are described in subsequent chapters. If cell sample concentrations are kept constant and acquisition is performed in a constant-time format, the absolute number of each gated population can be derived. Thus by comparing these numbers between the experimental and control samples, one can quantify specific cell death according to the following equations:

(a) Specific cell death (%) = (1 − live counts (7-AAD$^-$) of treated samples/live counts of mock-treated sample) × 100.

This reflects the % of viable cell loss by the experimental treatment, using 7-AAD exclusion alone.

(b) % Apoptosis = annexin V$^+$ counts/live counts (7-AAD$^-$) × 100 %.

This is a "snapshot in time" measurement for the fraction of cells undergoing early apoptosis (revealed by annexin V staining), before membrane integrity is compromised (7-AAD$^+$).

(c) % $\Delta\Psi$ loss = $DiOC_6$-low counts/live counts (7-AAD$^-$) × 100 %.

This represents a snapshot of cell fractions losing mitochondrial potential ($\Delta\Psi$) following treatment. Typically, loss of mitochondrial membrane potential occurs early in the apoptotic process, similar to phosphatidylserine exposure.

(d) Specific index of apoptosis = annexin V$^+$ counts of treated samples/annexin V$^+$ counts of the untreated control sample.

This calculation shows the change in apoptotic cells specifically induced by treatment. A similar index can be derived for cells losing ΔΨ by replacing the annexin V^+ counts with the $DiOC_6$-low counts.

4. Further Considerations for Data Interpretation

4.1. Percent of Cell Death Versus Viable Cell Loss

In most multicellular living organisms, PCD is a dynamic biological process that eventually leads to the removal of unwanted cells. In vivo, the clearance of cells undergoing PCD is rapid. As soon as phosphatidylserine (PS) is exposed on the membrane of apoptotic cells, it quickly triggers scavenger cells, including macrophages and other phagocytes, to engulf and destroy these dying cells. The fast clearance of apoptotic cells makes it extremely difficult to visualize and analyze the ongoing PCD process in vivo. For this reason, the majority of assays for apoptosis and other forms of PCD are carried out in vitro, seeking to partially recapitulate the in vivo process in the absence of scavenger cells. However, even in in vitro assays, many factors can vary such as the nature and strength of death stimulations, intrinsic preferential usage of PCD signaling pathways by different cells, and decomposition rates of dead cells—these differences will affect the dynamics of cell death processing, even in the absence of scavenger cells.

Taking a few Jurkat T cell lines for example, comparing the I2.1 line that has a deficiency in Fas-associated death domain (FADD), the 4E3 line that over-expresses tumor necrosis factor receptor-2 (TNFR2) only, and the 42.1 line that over-expresses TNFR2 plus a deficiency in FADD, their death responses to TNFα stimulation are significantly different. Under ideal conditions, assuming that the death and decomposition rates of the experimental cells were consistent, a snapshot of the percentage live or dead cells would be adequate for comparing effects of TNFα-induced death on these cells. However, the 42.1 cells are much more sensitive to TNFα-induced PCD as compared to that of I2.1 and 4E3 Jurkat cells (10). If we took a snapshot at 48 h post stimulation using a high dose (1 μg/ml) of TNFα, we might catch smaller differences for percentages of live or dead fractions among the three cell lines. Nevertheless, at the same time, one would likely also find that there might be tenfold more live 4E3 cells over live I42 cells after TNFα stimulation. Obviously, a simple percentage of live or dead cells, especially under high-dose death induction and at a later time point, may not be able to present fair comparisons for measuring the sensitivity of these different cell lines in response to death induction by TNFα.

To overcome the shortage of using % live/dead cells, the % cell loss calculation is recommended instead. Using TNFα induction

Table 2
Analyses of TNFα-induced PCD in Jurkat cells

Treatment	# Annexin V⁺ cells	# 7-AAD⁻ (live) cells	% Death
(A) RIP-deficient Jurkat T cells			
Mock	115	4,712	2.4
TNF-α	544	3,486	15.6
Index	4.7	74 % (26 % cell loss)	6.4
(B) FADD-deficient Jurkat T cells			
Mock	225	3,206	7
TNF-α	168	1,757	9.6
Index	0.7	55 % (45 % cell loss)	1.4

The numbers of annexin V+ and 7-AAD counts were collected from a flow-based kill assay experiment for induction of PCD with 10 ng/ml of TNFα for 24 h in RIP1-deficient (a) and FADD-deficient (b) Jurkat cells. The cells were gated from FCS/SSC similar to that shown in Fig. 2b

of cell death in Jurkat cells as an example, basically, a % cell loss is determined by comparing the absolute counts of live cells in the TNFα-treated against the mock-treated samples at a time point post TNFα stimulation. The investigators need to keep in mind that no matter what criteria are chosen for assessment of cell death (% dead cells, % loss of viable cells, apoptosis index, etc.), in an in vitro experiment, it can only be relatively accurate. This is because that at any given time point, a snapshot counting of cells is actually a function of accumulative proliferation of live cells and decomposition of dead cells. Table 2 depicts a scenario in which cell death indexes derived from comparing the absolute number of apoptotic cells (the bottom of column 2 for panels A and B) have significant fluctuations that are hard to interpret, whereas the % apoptotic cells or % cell loss (the bottom of column 3 for panels A and B) seems to be more representative of actual cell kill. For improving data presentation, an accurate calculation of % viable cell loss appears to be superior to other indices for most assessments of PCD in vitro kill assays.

4.2. Accurate Detection of Apoptotic Cells Using Annexin V

Annexin V is a Ca^{2+}-dependent phospholipid-binding protein with high affinity for PS, which normally exists inside of healthy cells. PS is exposed to the outside of the plasma membranes of cells undergoing apoptosis, even before membrane integrity damage happens; hence annexin V can be used to probe PS exposure during early apoptosis (5). However, annexin V cannot distinguish whether PS binding happens on the outside or the inside of a dying cell if it has lost membrane integrity, such as during necroptosis or at the late stages of apoptosis. Therefore, to detect apoptosis, annexin V must be used in conjunction with another dye that delivers a membrane integrity test, like PI or 7-AAD, in order to distinguish early apoptotic cells that have PS exposure yet retain undamaged membrane integrity. It is worth noting that the percentage of annexin V^+ and $PI^-/7\text{-}AAD^-$ apoptotic population in a given experiment may be low, very often below a double-digit level. It depends on how long the transition period lasts from the early to late stages of apoptosis that accompanies loss of membrane integrity.

In vivo, it has been shown that PS exposure quickly activates scavenger cells to phagocytose the dying cells, making the apoptotic cells disappear quickly (11, 12). However, since tissue macrophages and other scavengers are largely absent in most in vitro kill assay systems, the duration of apoptosis induction can vary appearing with some delay from the very first appearance of PS exposure until the loss of membrane integrity in apoptotic cells. This artificial delay allows investigators to visualize a distinct stage of apoptosis. However, the portion of early apoptotic cells (i.e., annexin V^+ and PI^-) can be 5 % or less, which poses difficulties for determination of data significance. In this situation, it is very helpful to acquire absolute numbers of live and dead cells, or in conjunction with other methods in support of a fair evaluation of the annexin V staining data.

4.3. Use of Time Courses and Dose Curves to Measure the Cell Death Process

It is well known that at the early stage of an immune response, antigen stimulation causes T lymphocytes to undergo activation and proliferation, while at the late stage, restimulation of the cycling T cells leads to immune contraction by PCD (1). Timing appears to be a critical factor for measuring the sensitivity of proliferating T cells to PCD inducing signals (13). For TCR RICD, freshly isolated T cells need to be activated through TCR and co-stimulatory signals for 2–3 days. The primary TCR stimulation needs to be removed by either washing out or extensively (5–10 times) diluting out the stimulants before adding in 50–100 IU/ml of interleukin-2 (IL-2) into the cultures. These IL-2-cultured T cells will become sensitive to RICD from day 5 and after. In general, mouse cycling T cells are faster and more sensitive than human T cells in becoming susceptible to RICD. Therefore, an RICD kill assay that spans 24, 48, and 72 h time points may be useful, especially when analyzing human T cells that tend to require more time for becoming RICD susceptible. Proper design of kill assays with appropriate timing may bolster your chances for detecting what are often subtle differences in a kill assay-based experiment. See Chapter 2 for a detailed protocol for the in vitro RICD assay using human T cells.

A multi-dose stimulation is also preferred for assessing susceptibility of lymphocytes toward RICD and death receptor (DR)-induced PCD. In our hands, using multi-dose titration experiments, deficiencies in RICD or DR-induced death of lymphocytes from autoimmune lympho-proliferative syndrome (ALPS) patients are detected mostly at low doses of TCR or DR stimulation conditions. Therefore, selecting appropriate time course and multiple dose titrations can greatly increase the precision as well as accreditation of RICD and DR-induced cell-death assays.

Unlike RICD, there is usually no need for a dose titration for measuring CWID of cycling T cells. Nevertheless, timing sufficient IL-2 culturing of activated lymphocytes appears to be critical; activated T cells need to be cultured in complete medium with high-dose IL-2 (100 IU/ml) for more than 3 days. It has been shown that these T cells stop making endogenous IL-2 shortly after the removal of TCR stimulation in vitro (14). In order to maintain T-cell growth, exogenous IL-2 needs to be supplemented in the culture at least every other day. These proliferating T cells can be cultured for more than a month, provided medium nutrients and IL-2 are regularly replenished. To set up a standard CWID assay, the IL-2-cultured T cells must be washed exhaustively with IL-2-free medium before setting a time course that covers a range of time points (e.g., 12, 24, 48, 72 h). Although investigators may utilize any of the aforementioned apoptosis detection methods, a time course measurement of viable cell loss is sufficient in studying CWID. A detailed protocol for the in vitro CWID assay is covered in Chapter 3.

4.4. Interpreting Cell Death Versus Suppression

The term "suppression" has been widely used and sometimes abused in immunology research. It often describes a scenario in which normal immune responses are "suppressed" based on measuring the inhibition of immune cell proliferation, cytokine production, and other immune functions. A puzzle exists here—namely, that the mechanisms underlying each case of immune suppression are actually elusive. A simple measurement of thymidine incorporation or cytokine production under particular experimental settings is actually insufficient to determine the underlying causes. For example, a decrease in cellular thymidine uptake or a drop in cytokine production in a given experiment can either be due to a decline in cell proliferation, and/or cytokine production, or a decrease of viable cell numbers that directly associate with the observation. For this reason, it is essential to include a kill assay that measures cell viability and death in addition to conventional proliferation assays and cytokine production assays, and thus to reach an impartial evaluation for the meaning of immune suppressions in each case.

4.5. Population Variance Affecting the PCD Sensitivity

It has been observed that $CD8^+$ T cells have significant survival advantage over $CD4^+$ T cells in vivo (15). We have observed a similar skewed growth pattern in cultured T lymphocytes in vitro, which favors the preferential survival of activated $CD8^+$ versus $CD4^+$ T cells over time (Zheng and Lenardo, unpublished data).

This might be one reason why, in kill assays using ex vivo activated primary T cells from different human donors, it is common to observe substantial differences in susceptibility to death induction by TCR or death receptor stimulations. This difference in susceptibility to death induction may be attributed to variable $CD4^+/CD8^+$ ratios, as well as contaminations of other cells in test samples. In order to avoid comparing "apples" with "oranges," it is recommended that a kill assay be carried out between 1 and 2 weeks post activation of isolated T cells whenever managing a mixture of CD4 and CD8 populations in culture. Alternatively, employing CD4/CD8 staining or using purified populations at the starting point of cell culture is preferred. It has also been suggested that the susceptibility of T cells to PCD induction may vary in different subpopulations, such as central versus effector memory T cells, NKT cells, etc. (see Chapter 4). Apparently, these factors also need to be considered for properly carrying out experiments and appropriate collection of meaningful data.

5. Conclusion

A kill assay based on membrane-dye exclusion is indisputable based on its simplicity and reliability for assessing viability and death of sample cells in vitro. It allows for rapid, efficient screening of apoptosis sensitivity in multiple samples. The flow cytometry-based kill assay is one of the most versatile methodologies available, capable of measuring the absolute number of live, dying, and dead cells with the option of labeling particular cell populations of interest. Subsequent chapters make use of this approach for studying RICD, CWID, and other PCD pathways.

Acknowledgements

This work was supported by the Intramural Research Program of the National Institute of Allergy and Infectious Diseases.

References

1. Snow AL et al (2010) The power and the promise of restimulation-induced cell death in human immune diseases. Immunol Rev 236: 68–82
2. Sampson J (1924) Determination of the resistance of leukocytes. Arch Intern Med 34:13
3. Boehme SA, Lenardo MJ (1992) Morphological, biochemical and flow cytometric assays of apoptosis. In: Coligan JE, Kruisbeek AM, Margulies DH, Shevach EM, Strober W (eds) Current protocols in immunology. Wiley, New York

4. Schmid I et al (1992) Dead cell discrimination with 7-amino-actinomycin D in combination with dual color immunofluorescence in single laser flow cytometry. Cytometry 13(2): 204–208
5. Vermes I et al (1995) A novel assay for apoptosis. Flow cytometric detection of phosphatidylserine expression on early apoptotic cells using fluorescein labelled Annexin V. J Immunol Methods 184(1):39–51
6. Packard BZ, Komoriya A (2008) Intracellular protease activation in apoptosis and cell-mediated cytotoxicity characterized by cell-permeable fluorogenic protease substrates. Cell Res 18(2):238–247
7. Mosmann T (1983) Rapid colorimetric assay for cellular growth and survival: application to proliferation and cytotoxicity assays. J Immunol Methods 65(1–2):55–63
8. Gavrieli Y, Sherman Y, Ben-Sasson SA (1992) Identification of programmed cell death in situ via specific labeling of nuclear DNA fragmentation. J Cell Biol 119(3):493–501
9. Koopman G et al (1994) Annexin V for flow cytometric detection of phosphatidylserine expression on B cells undergoing apoptosis. Blood 84(5):1415–1420
10. Chan FK et al (2003) A role for tumor necrosis factor receptor-2 and receptor-interacting protein in programmed necrosis and antiviral responses. J Biol Chem 278(51): 51613–51621
11. Fadok VA et al (1992) Exposure of phosphatidylserine on the surface of apoptotic lymphocytes triggers specific recognition and removal by macrophages. J Immunol 148(7): 2207–2216
12. Nagata S (2010) Apoptosis and autoimmune diseases. Ann N Y Acad Sci 1209:10–16
13. Peter ME et al (1997) Resistance of cultured peripheral T cells towards activation-induced cell death involves a lack of recruitment of FLICE (MACH/caspase 8) to the CD95 death-inducing signaling complex. Eur J Immunol 27(5):1207–1212
14. Zheng L et al (1998) T cell growth cytokines cause the superinduction of molecules mediating antigen-induced T lymphocyte death. J Immunol 160(2):763–769
15. Ferreira C et al (2000) Differential survival of naive CD4 and CD8 T cells. J Immunol 165(7):3689–3694

Chapter 2

Fluorescence-Activated Cell Sorting-Based Quantitation of T Cell Receptor Restimulation-Induced Cell Death in Activated, Primary Human T Cells

Gil Katz and Andrew L. Snow

Abstract

After initial stimulation with antigen and exposure to the growth cytokine interleukin-2, activated T lymphocytes become sensitized to apoptosis upon antigen restimulation through the T cell receptor. This self-regulatory, restimulation-induced cell death (RICD) program constrains the proliferative capacity of activated T cells to help prevent excessive T cell accumulation and associated immunopathology. Here we describe a simple FACS-based approach for measuring RICD sensitivity in activated human T cells following polyclonal restimulation in vitro. This procedure is a straightforward research and clinical diagnostic tool for assessing RICD sensitivity for T cells derived from normal donors and patients suffering from diseases causing dysregulated T cell homeostasis.

Key words: T cell receptor (TCR), Activated human T lymphocytes, IL-2, Anti-CD3, Fluoresence-activated cell sorting (FACS)

1. Introduction

Successful adaptive immune responses to infectious pathogens depend upon rapid clonal expansion of rare, antigen-specific T lymphocytes. Naïve T cells generally require three distinct signals to proliferate and acquire effector functions: (1) ligation of T cell receptor (TCR)/CD3 complexes via MHC-presented peptide antigen; (2) costimulation, primarily through the surface receptor CD28; and (3) exposure to the growth factor cytokine interleukin-2 (IL-2) in an autocrine/paracrine fashion (1). Generally, IL-2 induction of cell proliferation is a consequence of the first two signals, which upregulate expression of IL-2 as well as its high affinity receptor. Expansion of human peripheral blood T cells is readily induced in vitro using antibodies to nonspecifically cross-link CD3 and CD28 as well as exogenous, recombinant IL-2, as detailed

Andrew L. Snow and Michael J. Lenardo (eds.), *Immune Homeostasis: Methods and Protocols*, Methods in Molecular Biology, vol. 979, DOI 10.1007/978-1-62703-290-2_2, © Springer Science+Business Media New York 2013

in the Methods section below. In the context of infection, clonal T cell proliferation creates a potent arsenal of activated effector T cells that can directly eliminate pathogen-infected cells as well as provide assistance to B cells for effective humoral responses. An orderly transition from proliferation to contraction must take place as the adaptive immune response unfolds to eliminate excess T cells that may otherwise inflict unintended damage to host tissues, especially under conditions of recurring or persistent antigen (2). Programmed cell death via apoptosis both constrains and contracts the effector T cell pool throughout this period (3).

Restimulation-induced cell death (RICD) describes a propriocidal apoptosis mechanism by which reengagement of the TCR triggers death in a proportion of activated T cells (4). This phenomenon was originally described in vitro using mouse T cell hybridomas and/or primary, activated murine and human T cells cultured with exogenous IL-2, which readily undergo RICD upon restimulation with anti-CD3. Although the molecular details of this signaling pathway are not completely understood, RICD is proposed to act as a self-regulatory brake to set an upper limit for clonal expansion when antigen is abundant or recurrent (e.g., during chronic infections). Defective RICD can contribute to unchecked T cell accumulation and immunopathology in X-linked lymphoproliferative disease (XLP), underscoring the physiological relevance of this TCR-induced death program (5).

Here we provide a simple in vitro protocol for inducing and quantifying RICD sensitivity of activated human T cells using propidium iodide exclusion and flow cytometry. Complementary approaches for examining hallmarks of apoptosis during this process (e.g., Annexin V binding, caspase cleavage, mitochondrial depolarization, etc.) can easily be applied to the same anti-CD3-based restimulation procedure. This in vitro RICD assay provides a straightforward diagnostic tool for assessing death sensitivity in patients with aberrant T cell homeostasis, and allows basic and clinical researchers to investigate the genetic and biochemical mechanisms driving this death pathway.

2. Materials

2.1. Cell Isolation and Culture

1. Whole blood or buffy coat (see Note 1).
2. Ficoll-Paque PLUS (GE Healthcare).
3. 1× Phosphate-buffered saline (PBS).
4. 15 and 50 ml conical tubes; 1.5 ml microcentrifuge tubes (sterile).
5. ACK lysing buffer.

6. Complete RPMI medium: RPMI 1640 tissue culture medium (w/ 2–4 mM L-glutamine), 10% heat-inactivated fetal bovine serum (FBS), 100 U/ml penicillin–streptomycin. For 500 ml, add 50 ml FBS and 5 ml of 100× stock of penicillin–streptomycin. Store at 4°C.
7. Vented T75 tissue culture flasks (75 cm^2 surface area).
8. Anti-human CD3ε antibody (clone OKT3, Ortho Biotech, or clones HIT3a/UCHT1 (NA/LE), BD Biosciences).
9. Anti-human CD28 antibody (clone CD28.2 (NA/LE), BD Biosciences).
10. Recombinant human IL-2 (Peprotech or equivalent source).

2.2. RICD Assay and Flow Cytometry

1. 96-well round-bottom plate.
2. Propidium iodide (PI) (Sigma, 1 μg/ml stock diluted in 1× PBS).
3. 1.1 ml polypropylene microtubes (USA Scientific).
4. 5 ml round-bottom tubes for flow cytometry (BD Falcon, BD Biosciences).
5. Flow cytometry analyzer (FACScan, FACSCalibur, or equivalent).

3. Methods

3.1. Isolation of Peripheral Blood Mononuclear Cells Using Ficoll

1. Dilute the whole blood or buffy coat 1:2 with PBS (see Note 1). Distribute 20 ml of blood/PBS mixture per 50 ml conical tube.
2. Gently layer 12–13 ml of Ficoll under the blood/PBS mixture in each tube (see Note 2). Handle the tubes carefully to avoid disrupting the blood/Ficoll interface.
3. Spin the tubes at ~700 × *g* in a tabletop centrifuge for 20 min at room temperature (RT) with no brake (note that all subsequent centrifugations will be at RT).
4. After the spin, note the opaque layer of PBMC at the interface between the plasma (top, yellowish volume) and the Ficoll (clear). Carefully collect the PBMC from this interface using a pipette, and transfer to a new 50 ml conical tube (see Note 3).
5. Fill up to 50 ml with PBS and mix tubes by inverting several times.
6. Spin the tubes at 400 × *g* for 10 min to pellet PBMC. Carefully aspirate the PBS without disturbing pellet.
7. If excessive red blood cells (RBC) are visible in the pellet, resuspend the pellet in 5 ml ACK Lysing Buffer and mix gently by swirling the tube. Allow the cells to incubate at RT for 5 min, then fill up to 50 ml with PBS, and spin at 400 × *g* for 5 min to pellet.

8. Repeat the wash (steps 5–6) once more by resuspending PBMC pellet(s) in 5–10 ml PBS, then filling up to 50 ml PBS, and spinning at 400 × *g* for 5 min. Multiple pellets in separate tubes can be combined in one 50 ml wash.
9. Resuspend the cells in 30 ml of complete RPMI. Count PBMC using a hemocytometer.

3.2. Stimulation and Culture of Activated T Cells

1. Collect 10–20 million PBMC in a 50 ml conical tube and adjust the concentration to 2×10^6 cells/ml in complete RPMI.
2. Activate the T cells using 1 μg/ml each of anti-CD3 and anti-CD28 antibodies (1:1,000 dilution from 1 mg/ml stock). Add the antibodies directly to the tube, swirl to mix, and transfer to a T75 vented flask (see Note 4).
3. Incubate the cells in a humidified 37°C incubator (5% CO_2) for 3 days (see Note 5).
4. At ~72 h post activation, transfer cells to a 50 ml conical and pellet by spinning at 400 × *g* for 5 min. Wash the pellet twice by resuspending in ~30 ml PBS, inverting the tube several times, and spinning to pellet. Carefully aspirate PBS from the pellet after each wash.
5. After the second wash, resuspend pellet in 10 ml complete RPMI. Count cells and adjust concentration to $1–2 \times 10^6$ cells/ml.
6. Add exogenous recombinant human IL-2 to a final concentration of 100 international units (IU)/ml. Swirl to mix and transfer the volume to a fresh T75 flask.
7. Allow the cells to expand in culture for 7–10 days. Add or change media with fresh IL-2 (100 IU/ml) as needed every 2–3 days (see Note 6).

3.3. Plating T Cells for RICD Kill Assay

1. Count the T cells and collect 2×10^6 cells in a 15 ml conical tube. Pellet the cells by spinning at 400 × *g* for 5 min.
2. Aspirate the supernatant carefully and resuspend the pellet in 4 ml complete RPMI plus 100 IU/ml IL-2 (final cell concentration = 0.5×10^6 cells/ml).
3. Aliquot 0.5×10^6 cells (1 ml each) into 4 separate 1.5 ml microcentrifuge tubes.
4. Prepare diluted stocks of anti-CD3 mAb (clone OKT3, 1 mg/ml stock concentration) in complete RPMI plus 100 IU/ml IL-2, including 1, 10, and 100 μg/ml (see Note 7).
5. Add the dilutions of OKT3 mAb into each tube as described in Table 1 (see Note 8).
6. Mix by pipetting up and down several times, and then plate cells (1×10^5 cells/200 μl/well) in triplicate in a 96-well round-bottom plate (see Fig. 1).

Table 1
Recommended dilutions for OKT3 dose–response curve in RICD assay

Dilutions	Tube 1	Tube 2	Tube 3	Tube 4
1 μg/ml OKT3 diluted stock	–	10 μl	–	–
10 μg/ml OKT3 diluted stock	–	–	10 μl	–
100 μg/ml OKT3 diluted stock	–	–	–	10 μl
Final OKT3 concentration (ng/ml)	0	10	100	1,000

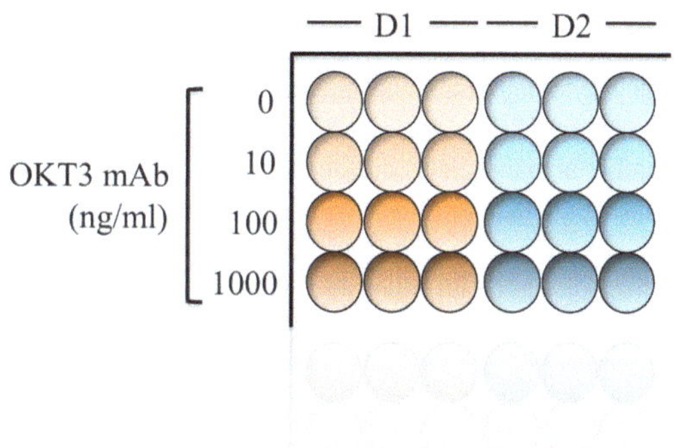

Fig. 1. Plating of activated human T cells for the RICD assay. Activated T cells (0.5×10^6 cells/ml) from different human donors (organized in columns; D1, D2, etc.) are treated with designated doses of OKT3 mAb (organized in *rows*) and subsequently plated in triplicate wells (1×10^5 cells/200 μl/well) for 24 h incubation. PI may be added directly to the wells before transferring cells to microtubes for FACS analysis.

7. Incubate the cells in a humidified 37°C incubator (5% CO_2) for 24 h to induce apoptosis (see Note 9).

3.4. FACS-Based Quantitation of RICD

1. At 24 h post restimulation, remove the plate from the incubator and stain cells by adding 10 μl of diluted PI stock (1 μg/ml) directly to each well (see Note 10).
2. Gently pipette up and down to mix and transfer all stained cells to racked 1.1 ml microtubes using a multichannel pipette.
3. Prepare the flow cytometer for data acquisition, including a 2-parameter dot plot for forward scatter (FSC) versus FL2 (PI channel). Note that the FSC scale is linear and FL2 is logarithmic.
4. To collect the cells, place each sample (cells in microtube) into a 5 ml FACS tube and vortex gently immediately prior to placement on the flow cytometer.

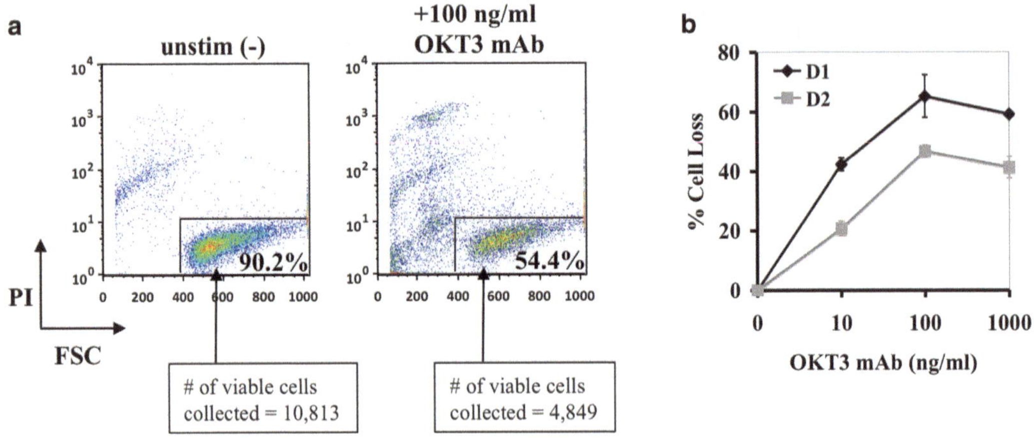

Fig. 2. Gating strategy for quantifying viable cells using PI exclusion. After restimulation (~24 h), PI-stained T cells are collected on the flow cytometer for a constant time (20–30 s/sample). (**a**) For analysis, data is visualized in a 2-color dot plot with forward scatter (FSC) on the *x*-axis and PI (FL2) on the *y*-axis. Viable cells (FSC^{hi}, PI-) are gated as shown to determine the number of viable cells collected per sample. Percent cell loss is then calculated as follows:
# of viable cells (20 sec collection): unstim (0 ng/ml) = 10813; + 100 ng/ml OKT3 = 4849; % cell loss = [1-(4849/10813)] * 100 = 55.2%
(**b**) By repeating this calculation for triplicate values, a dose–response curve for mean RICD sensitivity (–/+ SD) can be readily generated to compare different donors (*Panel B*).

5. Collect each sample using constant time (20–30 s/sample) (see Note 11).
6. For analysis, use FACS analysis software (e.g., FlowJo, FACSDiva, or equivalent) to plot FSC versus FL2 for each sample, gating on the viable cells (FSC^{hi}/PI^{-} population, see Fig. 2) to determine the number of viable cells collected.
7. Calculate the % cell loss for each dose of OKT3 as follows: % cell loss = [1 – (# of viable (PI-) restimulated cells/# of untreated viable cells)] × 100, see Note 12.
8. Plot % cell loss data on a dose–response curve (see Fig. 2 for example).

4. Notes

1. For best PBMC viability, whole blood may be collected in tubes with acid citrate dextrose (ACD) as an anticoagulant. Alternatively, the enriched PBMC fraction obtained from a whole unit blood donation after initial centrifugation (a.k.a. the "buffy coat") may be used to obtain large quantities of PBMC.
2. For best results to layer Ficoll underneath blood/PBS mixture, we recommend placing a sterile 5.75 in. glass Pasteur pipette

and adding up to 3 ml Ficoll to the top of the pipette to allow Ficoll to drip slowly from the narrow opening at the bottom of the tube. Repeat this step until 13 ml of Ficoll is distributed per tube (20 ml blood/PBS). Alternatively, blood/PBS mix can be slowly layered on top of pre-aliquoted Ficoll using a sterile pipette, being careful not to disrupt the interface.

3. The volume of PBMC transferred to each 50 ml conical tube should not exceed 8 ml, as too much residual Ficoll may interfere with pelleting of cells by centrifugation and reduce yield. For best viability of isolated PBMC, cells must be washed thoroughly to remove all Ficoll.

4. Use of soluble anti-CD3/CD28 antibodies should activate T cells efficiently in the context of all PBMC, as the presence of Fc receptor-bearing "accessory cells" (i.e., monocytes, B cells) will aid in cross-linking. Other polyclonal stimuli may also be utilized to activate T cells in this context (e.g., concanavalin A (5 μg/ml) or phytohemagglutinin (PHA)-L (1–2 μg/ml), both available from Sigma). If T cell subsets are purified from PBMC prior to initial stimulation, we recommend using plate-bound anti-CD3/CD28 Abs or Ab-coupled beads (e.g., T cell activation/expansion kit, Miltenyi Biotec).

5. Cells should be monitored for signs of proper T cell activation during this period. T cells should blast (i.e., increase in size) and form small clumps in culture by day 2–3. Alternatively, the upregulation of certain cell surface markers (e.g., CD69 on day 1, CD25 on days 2–3) may also be tracked using flow cytometry.

6. The addition of exogenous IL-2 should result in rapid cell division and expansion of T cells in the first 7–10 days. The T cells will be largely comprising activated $CD4^+$ and $CD8^+$ cells, with the proportion of $CD8^+$ T cells increasing over time. IL-2 is always kept in excess in the culture to sensitize cells to RICD (6) and avoid inducing cytokine withdrawal-induced death (see Chapter 3). Monitor cultures every 1–2 days by watching media color (i.e., red to yellow color change of RPMI containing phenol-red accompanies the active growth and proliferation of cells since they produce metabolic acids) and counting cells. Ideally, cultures should be kept at a density of $1–2 \times 10^6$ cells/ml throughout this period. Feed or split cells accordingly using complete RPMI and fresh IL-2.

7. The anti-CD3 Ab clone OKT3 should reliably induce apoptosis in ~40–60% of activated T cells (~100 ng/ml), although sensitivity will vary depending on the donor and the relative proportion of $CD8^+$ T cells (~70–80% at the time of assay). Purification of $CD4^+$ or $CD8^+$ T cells prior to restimulation may be done to assess subset-specific RICD sensitivity and reduce donor-dependent variability. For assaying patients with

suspected RICD defects, we recommend utilizing at least 2 separate controls activated and cultured in parallel. Other anti-CD3 antibodies will also induce RICD with variable potency (e.g., clone 64.1 > OKT3; clone HIT3a ≈ OKT3). In our hands, the addition of anti-CD28 Ab does not affect the results of the RICD assay.

8. The recommended dose range of OKT3 mAb (0.01–1 μg/ml) usually encompasses maximum RICD sensitivity in normal donor T cells, typically induced at ~100–200 ng/ml. Donor-dependent differences in RICD sensitivity are often more readily observed using lower doses of OKT3, which may necessitate the addition of lower dilutions (1–5 ng/ml) to the dose–response curve. Plate-bound anti-CD3 will produce a more robust restimulation signal and increased apoptosis at all doses—while this is optional for assessing RICD sensitivity in human cells, it is required for inducing apoptosis in mouse T cells.
9. The majority of RICD-sensitive cells will die by apoptosis (i.e., become PI^{+}) by ~18 h post restimulation. We routinely assay cells by PI exclusion ~20–24 h post restimulation with little change in % cell loss.
10. PI rapidly intercalates into the double-stranded DNA of apoptotic cells that have lost plasma membrane integrity. Similar intercalating dyes (e.g., 7-AAD) may be substituted for PI. Moreover, earlier markers of apoptosis (e.g., Annexin V staining, cleavage of caspase-3, loss of mitochondrial membrane potential) can be measured ~4–8 h post restimulation by flow cytometry using the same basic restimulation setup.
11. Samples are collected on constant time for proper calculation of % cell loss (see Note 12). We advise waiting 2–3 s after placing each tube on the flow cytometer before starting acquisition. This allows the flow rate to stabilize and improves consistency in collecting a similar number of events (>10,000 cells) per sample.
12. By counting the *number* of viable cells collected for the same amount of time from each sample, the % cell loss calculation accounts for cells that disintegrate after apoptosis and do not register as events on the flow cytometer. This calculation therefore provides greater sensitivity than direct measurement of % PI^{+} dead cells, although % dead cells and % cell loss should be similar.

Acknowledgements

This work was supported by grants from Uniformed Services University and the XLP Research Trust.

References

1. Smith-Garvin JE, Koretzky GA, Jordan MS (2009) T cell activation. Annu Rev Immunol 27:591–619
2. Arnold R, Brenner D, Becker M, Frey CR, Krammer PH (2006) How T lymphocytes switch between life and death. Eur J Immunol 36:1654–1658
3. Bidere N, Su HC, Lenardo MJ (2006) Genetic disorders of programmed cell death in the immune system. Annu Rev Immunol 24:321–352
4. Snow AL, Pandiyan P, Zheng L, Krummey SM, Lenardo MJ (2010) The power and the promise of restimulation-induced cell death in human immune diseases. Immunol Rev 236: 68–82
5. Snow AL, Marsh RA, Krummey SM, Roehrs P, Young LR, Zhang K, van Hoff J, Dhar D, Nichols KE, Filipovich AH, Su HC, Bleesing JJ, Lenardo MJ (2009) Restimulation-induced apoptosis of T cells is impaired in patients with X-linked lymphoproliferative disease caused by SAP deficiency. J Clin Invest 119:2976–2989
6. Lenardo MJ (1991) Interleukin-2 programs mouse alpha beta T lymphocytes for apoptosis. Nature 353:858–861

Chapter 3

Evaluation of IL-2-Withdrawal-Induced Apoptosis in Human T Lymphocytes

Joao Bosco Oliveira

Abstract

Proper lymphocyte apoptosis is critical for the maintenance of immune system homeostasis, and evaluation of cell death is useful in a number of clinical and research settings. We describe here how to evaluate the integrity of the intrinsic pathway of lymphocyte apoptosis triggered by starvation for IL-2 or other survival cytokines.

Key words: Apoptosis, Lymphocytes, Intrinsic pathway, IL-2 starvation

1. Introduction

Maintenance of lymphocyte homeostasis by elimination of autoreactive, damaged, or senescent lymphocytes is important for the prevention of autoimmunity and cancer (1). Two main pathways mediate lymphocyte apoptosis: the extrinsic and the intrinsic or mitochondrial pathways (1). The extrinsic pathway is modulated by the interaction of TNF-superfamily surface receptors and their ligands, having FAS/CD-95/APO-1 as its prototype (2). Disruptions of the FAS pathway in humans are associated with the development of the autoimmune lymphoproliferative syndrome (ALPS), characterized by autoimmunity, lymphocyte accumulation, and an increased lymphoma risk, illustrating its importance for human lymphocyte homeostasis (3). On the other hand, the intrinsic pathway of apoptosis involves a mitochondrial pathway controlled by the BCL-2 family of proteins, which has anti- and pro-apoptotic members (4). Somatic mutations in molecules of the intrinsic pathway are classically associated with tumor development, but more recently were also associated with the appearance of a syndrome, Ras-associated autoimmune leukoproliferative

Andrew L. Snow and Michael J. Lenardo (eds.), *Immune Homeostasis: Methods and Protocols*, Methods in Molecular Biology, vol. 979, DOI 10.1007/978-1-62703-290-2_3, © Springer Science+Business Media New York 2013

disorder (RALD) involving autoimmunity and accumulation of lymphocytes and monocytes (5, 6).

For the disorders listed above and many others, evaluation of apoptosis is an integral part of the cellular diagnostic workup. Although assays for checking FAS function in human lymphocytes are more commonly described in the literature, detailed methodological description on how to evaluate the intrinsic pathway is lacking. In this chapter we describe how to assess human lymphocyte apoptosis after triggering the intrinsic pathway by IL-2 starvation, mimicking the end phase of an immune response (7).

2. Materials

2.1. Cell Culture

1. Peripheral blood lymphocyte isolation: 1.077 g/ml Ficoll-Hypaque, 1× PBS, pH 7.4. Store at room temperature.
2. Cell culture medium: RPMI 1640, 2 μM L-glutamine, 100 IU/ml penicillin, 100 μg/ml streptomycin, 10 % fetal bovine serum (see Note 1). To prepare 500 ml of cell culture media mix 440 ml of RPMI 1640 with 5 ml of 0.2 M L-glutamine, one 10 ml aliquot of 100× penicillin/streptomycin, and 50 ml of fetal bovine serum. Sterile filter the mixture (0.22 μm membrane) and store the bottle at 4 °C for up to 6 weeks.
3. T lymphocyte activation and growth: Recombinant human IL-2 (Roche Applied Science), activating anti-CD3 monoclonal antibody (OKT3, Ortho Biotech) (see Note 2).

2.2. Apoptosis Measurement

1. 40 nM $DiOC_6(3)$ (3,3′-dihexyloxacarbocyanine iodide). Prepare a 0.1 mM solution of $DiOC_6$ (3) (Molecular Probes) by dissolving 5.7 mg dye in 10 ml dimethyl sulfoxide (DMSO). Store for 3 months in small (e.g., 0.5 or 1 ml) aliquots protected from light at −20 °C. Prior to use, dilute tenfold with PBS to obtain 10 μM working aliquots.
2. 1 μg/ml Propidium iodide. To make a stock solution dissolve PI in deionized water (dH_2O) at 1 mg/mL (1.5 mM) and store at 2–6 °C, protected from light, for up to 6 months.

3. Methods

3.1. Lymphocyte Culture and Activation

Carry all procedures at room temperature and under sterile conditions, unless noted otherwise.

1. To separate PBMCs from whole blood or buffy coats overlay 20 ml of a 1:1 mixture of blood:PBS over 7 ml of 1.077 Ficoll

and centrifuge at 300 × *g* for 20 min at room temperature (RT) (see Note 3). Remove the white PBMC layer in the plasma/Ficoll interface, transfer to a 50 ml polypropylene conical tube, and wash the cells in PBS. Centrifuge for 5 min at 200 × *g*, discard the supernatant, resuspend the cells in complete culture medium, and count.

2. Adjust cell concentration to 1×10^6 lymphocytes/ml and add IL-2 to a final concentration of 25 international units (IU)/ml plus 1 μg/ml OKT3. Incubate at 37 °C, 5 % CO_2 for 3 days on a cell culture flask of adequate size (see Note 4).
3. At the end of the 3-day activation and proliferation period, transfer the cells to a 50 ml conical (in the laminar flow hood) and centrifuge at 200 × *g* for 5 min. Discard the old medium, wash cells twice with PBS, and resuspend the cells in fresh culture medium supplemented with 100 IU/ml IL-2. No OKT3 should be added at this point. Adjust the cell concentration to 1×10^6 cells/ml and culture for 3 more days at 37 °C, 5 % CO_2.
4. At the end of the 6th day, transfer cells to a 50 ml conical tube and centrifuge at 200 × *g* for 5 min. Count cells and resuspend at 1×10^6 cells/ml on fresh culture media supplemented with 100 IU/ml IL-2. The apoptosis experiments can begin on the next day (see Note 5).

3.2. Apoptosis Induction and Evaluation

1. To starve cells from IL-2, remove the cells from culture, transfer to a conical tube, and wash them 3 times in PBS. After each wash, centrifuge cells at 200 × *g* for 5 min. Resuspend the cells in fresh culture medium without IL-2 at 1×10^6 cells/ml, and then plate 200 μl (200,000 cells)/well in triplicates in round-bottom 96-well plates. Plate enough cells for at least 5 days of analysis (see Notes 6 and 7).
2. Measure apoptosis every day (starting from day 0) for 5 consecutive days. For the apoptosis measurement, remove the plate from the incubator and add PI and $DiOC_6$ to a final concentration of 1 μg/ml of propidium iodide (PI) and 40 nM of $DiOC_6$. Incubate the plate for 15 min at 37 °C. No washes are necessary. After the 15-min incubation time transfer the cells to pre-labeled flow cytometry 12 × 75 mm tubes and take them to the flow cytometer for data acquisition (see Note 8).
3. Live cells (PI negative and $DiOC_6$ positive) are collected by flow cytometry using a constant time acquisition (see Notes 9 and 10). The PI signal can be acquired on the FL2 channel, and $DIOC_6$ on the FL1 channel. Remember to have an unstained tube (for calibration) and one tube with only PI and another with only $DIOC_6$, for fluorescence compensation (Fig. 1) (see Note 11).

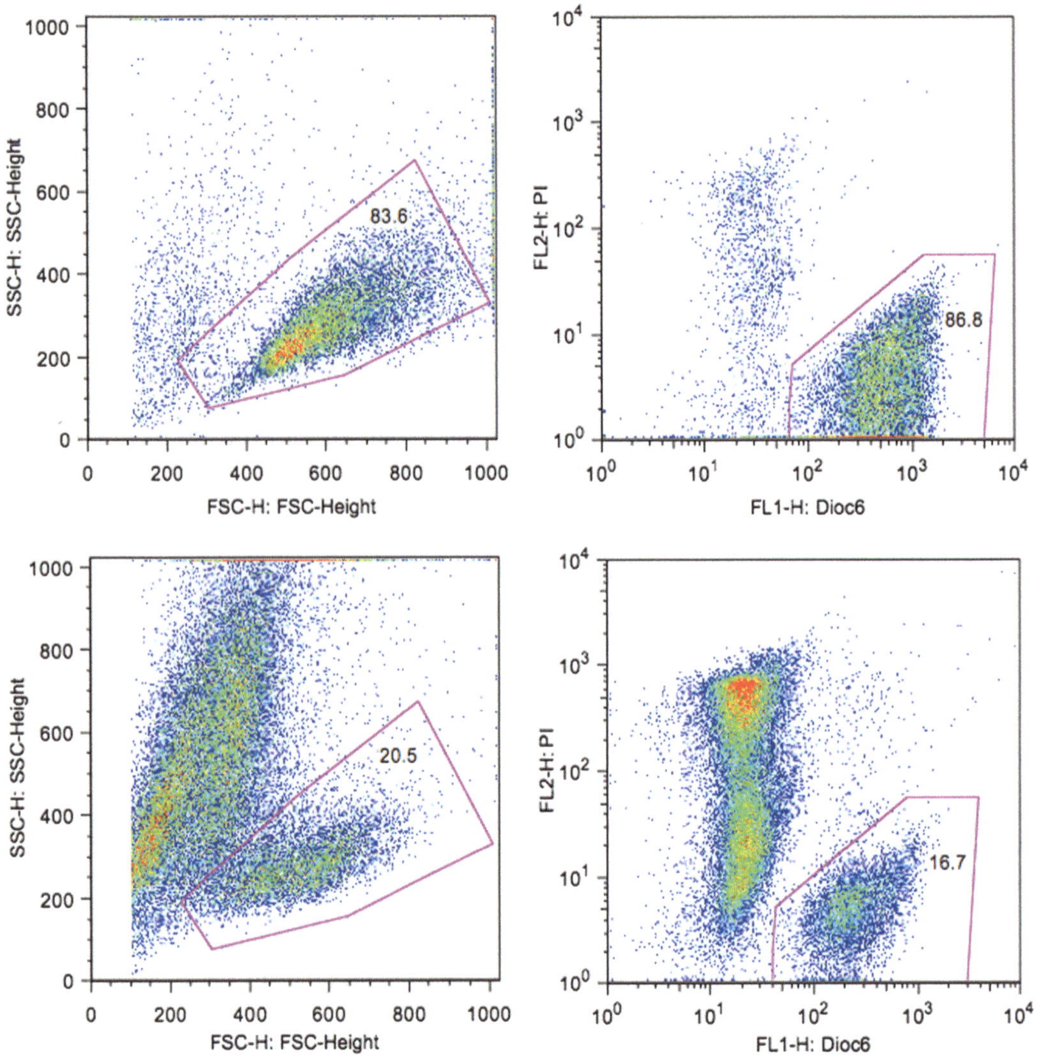

Fig. 1. Example of data collection by flow cytometry. The upper plots were drawn from data collected at the beginning of the assay. The number of live cells is determined by gating around DIOC$_6^+$PI$^-$ cells. This number usually corresponds to the number of live cells as determined by FSC × SSC plots (*left panel*). Live cells are typically defined as FSC^{hi}SSC^{lo}. The lower plots illustrate the typical results seen at 72 h after IL-2 withdrawal, with fewer cells in the DIOC$_6^+$PI$^-$ gate, and many cells moving into the DIOC$_6^{low}$PI$^-$ (early apoptosis) and DIOC$_6^{low}$PI$^-$ gate (late, necrotic cells). Please note also the decrease of live cells using the FSC × SSC gate (*left panel*).

3.3. Data Analysis

1. The amount of cell loss is calculated according to the following formula: ([number of live cells at 0 h − number of live cells at later time points]/number of live cells at 0 h) × 100 (see Notes 12 and 13). It is not uncommon to notice an increase in the number of live cells 24 h after IL-2 starvation, as many cells were in the middle of a cycle when starved. A significant decrease in viability is typically noted after 72 h of starvation (Fig. 2).

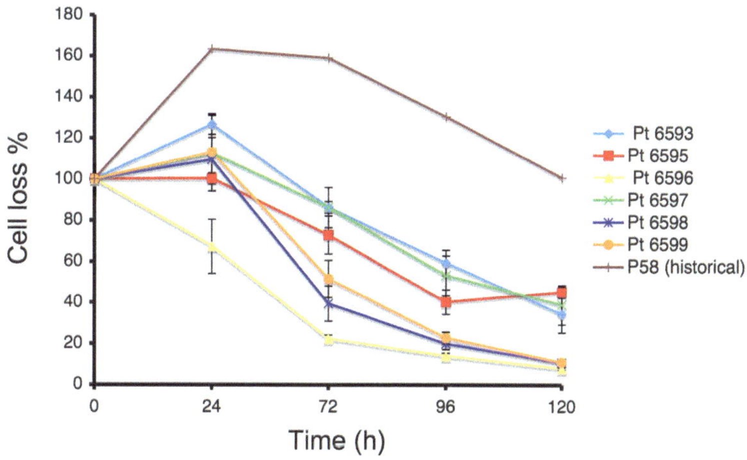

Fig. 2. Typical results after IL-2 starvation of T cells. Plotted here are several normal donors and one patient (P58) with an intrinsic pathway defect secondary to a gain-of-function mutation in *NRAS*. Note the increased viability of this sample at 120 h, as compared to the controls.

4. Notes

1. Mix 5 ml of the 100× penicillin stock with 5 ml of 100× streptomycin stock and store the 10 ml aliquots at −20 °C, to make your life easier.
2. Keep the IL-2 stock solution at −70 °C, in small aliquots, to avoid repeated freezes and thaws. Store working aliquots at 4 °C for 1 month.
3. Separating PBMCs using Ficoll must be done at room temperature to ensure proper separation of blood components. In addition, you must turn off the rotor brake to allow a slow decline in speed. Otherwise the sudden deceleration can disrupt the layers.
4. Make sure that the CO_2 levels in your incubator are calibrated, and that the water pan is full.
5. Cells have to be in good shape to conduct IL-2 withdrawal apoptosis experiments. For human cells, this means using them from the 7th to the 14th day in culture. After this time point, intracellular levels of pro-apoptotic proteins of the BCL-2 family (especially BIM) are high, leading to erratic results. Also, it is critical to keep the cells well fed while in culture. Count the cells and adjust to 1×10^6 cells/ml every 2 or 3 days, and replace fresh medium with IL-2.
6. Be gentle with the cells. After each wash, resuspend them from the bottom of the tube by tapping on the tube, not by pipetting up and down. Also, avoid centrifuging at higher speeds.

7. If you are short on cells, you can decrease the number of cells per well to 50,000. However, it is better to maintain the 200 μl final volume, as the assay takes several days and using less volume can lead to excessive evaporation and variable results.
8. To facilitate staining with PI and $DiOC_6$, dilute and mix them together at 20× final concentration right before adding to the plate, and use a repeat pipettor to add 10 μl to each well. This can save you a lot of time, especially if you have many samples.
9. Because cell loss is measured over time by counting viable cells on the flow cytometer, it is critical that the wells do not lose volume during the incubation period. To minimize evaporation, make sure that the incubator water pan is full and add dH_2O to all empty wells on your plate. Additionally, if you do not have many samples, plate them in the middle of the plate, surrounded by wells with dH_2O.
10. For 200,000 cells/well, usually 15 s/tube is a good acquisition time, on high-speed mode. If you are using less cells make sure to use a time long enough to allow acquisition of at least 10,000 events/tube. Do not dilute the samples before acquisition, as the volume and consequently the number of events acquired in the selected amount of time will vary. Make sure that there is no dripping from your flow cytometer into the sample tube. Newer flow cytometer instruments do not allow sheath fluid to drip into the sample, but older models might.
11. $DiOC_6$ fluorescence intensity can look very bright, with significant bleeding into the FL2 channel. If this is the case, dilute the $DiOC_6$ further and perform proper compensation in a subsequent experiment.
12. As this is a very prolonged assay, we prefer to measure the loss of live cells rather than the presence of dead cells, as apoptotic cells may become necrotic, fragment, and disappear as they become smaller than the preset size threshold. However, as we are collecting data using DiOC6 and PI staining, one can easily reanalyze the data based on percentage of dead cells at each time point.

References

1. Lenardo M, Chan KM, Hornung F et al (1999) Mature T lymphocyte apoptosis-immune regulation in a dynamic and unpredictable antigenic environment. Annu Rev Immunol 17: 221–253
2. Tibbetts MD, Zheng L, Lenardo MJ (2003) The death effector domain protein family: regulators of cellular homeostasis. Nat Immunol 4(5):404–409
3. Bidere N, Su HC, Lenardo MJ (2006) Genetic disorders of programmed cell death in the immune system. Annu Rev Immunol 24: 321–352
4. Strasser A (2005) The role of BH3-only proteins in the immune system. Nat Rev Immunol 5(3):189–200
5. Oliveira JB, Bidere N, Niemela JE et al (2007) NRAS mutation causes a human autoimmune

lymphoproliferative syndrome. Proc Natl Acad Sci U S A 104(21):8953–8958

6. Niemela JE, Lu L, Fleisher TA et al. Somatic KRAS mutations associated with a human non-malignant syndrome of autoimmunity and abnormal leukocyte homeostasis. Blood. 10; 117(10):2883–2886.
7. Strasser A, Pellegrini M (2004) T-lymphocyte death during shutdown of an immune response. Trends Immunol 25(11):610–615

Chapter 4

Determination of Apoptosis Sensitivity in Specific T Cell Subsets from Human Peripheral Blood by Utilizing a Multiparameter Fluorescence-Activated Cell Sorting-Based Technique

Bernice Lo and Madhu Ramaswamy

Abstract

Among the different techniques available for determining physiological cell death or apoptosis in immune cells, fluorescence-activated cell sorting-based approaches prove to be one of the most efficient and quantitative assays in capturing cells that are actively undergoing apoptosis elicited by either extrinsic or intrinsic forms of cell death. The key advantage of the technique is to allow the user to determine apoptotic responses of multiple immune cell types or subsets of the same lineage of immune cells without the prior requirement for cell separation. Here, we describe a "multiparameter" flow method to rapidly determine apoptosis-sensitivity induced by the Fas receptor pathway within different CD4 T cell subsets in a mixed pool of analyzed ex vivo peripheral blood mononuclear cells.

Key words: Fas, Apoptosis, Multi-parameter, flow cytometry, Ex vivo, annexin, T cell subsets, Effector memory (T_{EM}), Central memory (T_{CM}), Naïve CD4+ T cells, Memory CD4+ T cells, Autoimmunity, Autoimmune Lymphoproliferative Syndrome (ALPS)

1. Introduction

$CD4^+$ T cell apoptosis mediated by the Fas/CD95/APO-1 receptor/ligand pathway is critical in establishing peripheral self-tolerance (1). Deficiency or mutations of the Fas receptor in mice or humans result in the accumulation of autoreactive $CD4^+$ T cells with overt systemic autoimmunity (2–4). Recent work indicates that Fas does not induce the elimination of all activated $CD4^+$ T cells, but selectively targets the effector memory (T_{EM}) subset of the memory $CD4^+$ T cell pool in ex vivo derived peripheral blood mononuclear cells (PBMCs). The naïve and central memory (T_{CM})

Andrew L. Snow and Michael J. Lenardo (eds.), *Immune Homeostasis: Methods and Protocols*, Methods in Molecular Biology, vol. 979, DOI 10.1007/978-1-62703-290-2_4, © Springer Science+Business Media New York 2013

subsets, on the other hand, show reduced sensitivity to Fas-induced apoptosis (5). This finding correlates well with the known functional requirement of Fas apoptosis in curtailing immunopathology mediated by autoreactive T cells that may possibly originate from T_{EM} cells, without affecting acute antigenic or recall CD4$^+$ T cell responses (6). Since only the T_{EM} subsets within the CD4$^+$ T cell pool are specifically sensitive to Fas-induced apoptosis, it is essential to introduce a multiparametric gating component when testing apoptosis defects in a mixed immune cell pool of patient PBMCs. Importantly, such a multiparameter flow cytometry apoptosis assay can be useful in clinical testing for autoimmune lymphoproliferative syndrome (ALPS), a disease caused by defective Fas-mediated apoptosis (7, 8), or other autoimmune diseases associated with defective T cell apoptosis phenotypes (9). Current apoptosis assays are labor intensive and time consuming, since they require the activation of T lymphocytes and extended culture in interleukin-2 (IL-2), prior to testing for Fas-induced apoptosis (10). In addition, differences in activation and prolonged culture conditions can lead to inconsistent test results. Thus, the rapid ex vivo apoptosis assay described here eliminates some of these problems and also provides a more convenient assay for clinical use. The following ex vivo multiparameter apoptosis protocol describes detailed steps to assay for Fas-induced apoptosis in different subsets of CD4$^+$ T cells within a pool of donor PBMCs using flow cytometry.

2. Materials

1. Whole blood (see Note 1).
2. Ficoll-Paque™ PLUS (GE Healthcare).
3. 1× Phosphate-buffered saline (PBS).
4. 50 mL polypropylene conical tubes.
5. ACK lysing buffer.
6. Complete RPMI culture medium: RPMI 1640 tissue culture medium with 2 mM L-glutamine, 10 % heat-inactivated fetal bovine serum (FBS), 100 U/ml penicillin–streptomycin. For 500 ml, add 50 ml heat-inactivated FBS and 5 ml of 100× stock of penicillin–streptomycin. Store at 4 °C.
7. 96-well flat-bottom plate.
8. Anti-Fas antibody (clone APO-1.3, Enzo Sciences).
9. Anti-IgG_3 antibody.
10. Annexin V binding buffer (ABB) with bovine serum albumin (BSA): 0.5 % BSA, 10 mM HEPES, 140 mM NaCl, and 2.5 mM $CaCl_2$, pH 7.4. Store at 4 °C.

3. Methods

Carry out all procedures at room temperature unless otherwise specified.

3.1. Ficoll Separation of Peripheral Blood Mononuclear Cells

1. Dilute whole blood 1:2 with PBS (see Note 1). Aliquot 20 mL of blood/PBS mix per 50 mL conical and layer 13 mL of Ficoll under the blood (see Note 2).
2. Spin at ~700 × *g* for 20 min with no brake at room temperature (RT).
3. Using a transfer pipette, gently collect the PBMCs from the interface between the plasma and Ficoll layers into a 50 mL conical. Minimize the amount of Ficoll collected along with the cells.
4. Add volume up to 50 mL with PBS to wash. Spin down cells at ~400 × *g* for 10 min. Aspirate to remove the PBS.
5. Wash 2 more times with 40–50 mL of PBS (see Note 3). Spin at ~400 × *g* for 10 min for each wash and aspirate to remove PBS.
6. If there are RBCs visible, resuspend the pellet in 5 mL of ACK lysing buffer and swirl for 1–2 min at RT. Bring volume up to 50 mL with PBS to wash as above. Wash once more and aspirate to remove PBS.
7. Resuspend cells in 10 mL of complete RPMI.

3.2. Plating the Ex Vivo Apoptosis Assay

1. Prepare cross-linked anti-Fas antibody (APO-1.3) dilution(s) (see Note 4). Use anti-IgG_3 to cross-link anti-Fas (see Note 5). For more sensitive detection of apoptosis differences, a dose–response curve is recommended. Suggested concentrations are 1,000 ng/mL, 100 ng/mL, and 10 ng/mL of APO-1.3 at 2× working concentration (Fig. 1).
2. For each APO-1.3 dilution, aliquot 100 μL per well to 3 wells of a 96-well flat-bottom plate. Aliquot 100 μL per well of complete RPMI alone as a negative control.
3. Plate 2–3 additional wells with 100 μL of APO-1.3 at 2,000 ng/mL to be used later for single-color compensation (see Note 6).
4. Count cells and dilute cells to 2×10^6 cells/mL in complete RPMI. Aliquot 100 μL of cells to each well plated above. Save 1–2 million cells for single-color compensation staining.
5. Incubate in a CO_2 incubator at 37 °C for 6–8 h (see Note 7).

3.3. Post-apoptosis Assay Processing: Multiparameter FACS Staining

1. After 6–8 h of incubation with the apoptosis agent, transfer cells from the flat-bottom 96-well plate to a round-bottom plate in order to stain them for surface and apoptosis markers (see Note 8).

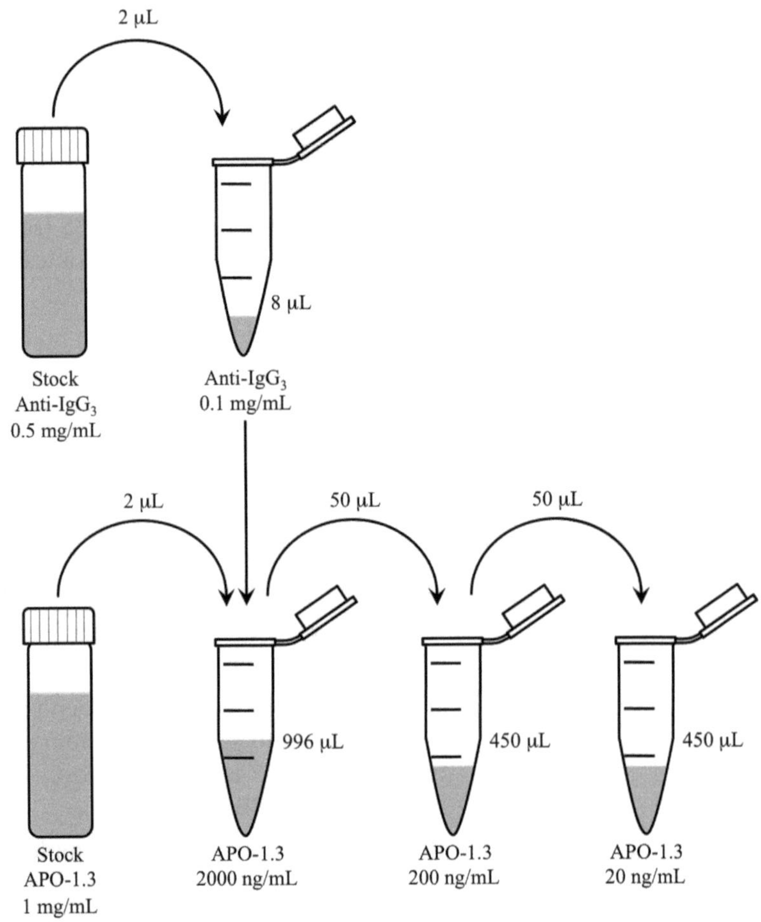

Fig. 1. Schematic diagram of serial dilution of antibodies.

2. Wash the cells twice with ABB (see Note 9) by adding 200 µL of buffer into each well and spinning the plates at ~400 × *g* for 5 min.
3. Prepare the staining mix for simultaneous staining of T cell surface markers along with annexin V in ABB and add 30 µL of staining mix per well (2×10^5 cells). We recommend the following surface markers to best identify the Fas-sensitive effector memory subsets within a mixed pool of PBMCs: CCR7, CD45RA, CD4, annexin V, and CD27 (see Note 10).
4. Stain for at least 20 min on ice and wash cells twice with ABB before proceeding to fix them for 10 min with 3 % paraformaldehyde at room temperature (see Note 11).
5. Wash the fixed cells twice with fluorescence-activated cell sorting (FACS) buffer, resuspend in 100 µL of buffer, and proceed to FACS analysis (see Note 12).

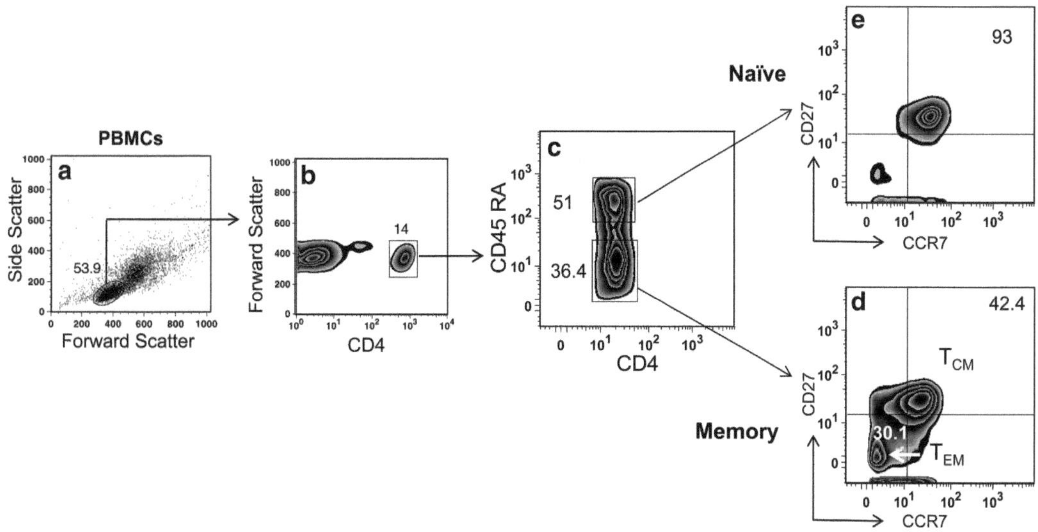

Fig. 2. Gating strategy for analysis of naïve and memory $CD4^+$ T cell subsets within a mixed population of PBMCs. Viable (FSC^{hi}, SSC^{lo} as in **a**) $CD4^+$ bright (as in **b**) T cells are gated from within an initial lymphocyte population on a mixed pool of PBMCs as indicated. CD45RA surface marker further subdivides naïve ($CD4^+CD45RA^+$) and memory ($CD4^+CD45RA^-$) CD4 T cells and presence/absence of CCR7 and CD27 differentiates between T_{CM} ($CCR7^+/CD27^+$) and T_{EM} ($CCR7^-/CD27^-$) (**d, e**).

3.4. Gating Strategy for Data Acquisition and Calculation of Cell Death

1. Gate on the lymphocyte population within the mixed PBMC pool on FSC versus SSC (Fig. 2a).
2. Within the lymphocyte population, gate for $CD4^+$ T cells. The marker CD45RA differentiates naïve ($CD4^+CD45RA^+$) and memory ($CD4^+CD45RA^-$) $CD4^+$ T cells (Fig. 2c) (see Note 13).
3. Further subsets within the memory $CD4^+$ T cell pool are delineated using the chemokine receptor CCR7 and the TNF superfamily receptor CD27 (Fig. 2d). As indicated in the figure, central memory (T_{CM}) subsets are defined by $CCR7^+/CD27^+$ staining cells, whereas $CCR7^-/CD27^-$ cells fall under the effector memory (T_{EM}) subsets. Naïve $CD4^+$ T cells, on the other hand, tend to stain double positive for both the markers ($CCR7^+/CD27^+$) (Fig. 2e), similar to T_{CM} cells (see Note 14).
4. We recommend collecting at least 10,000 (5,000 at minimum) gated events for each subset to have significant cell numbers for the final calculation of cell death based on annexin V staining.
5. Post-FACS analysis: To analyze the extent of Fas-induced apoptosis within each subset, FACS software such as FlowJo is used to assess annexin staining within each $CD4^+$ T cell subset. In order to account for dead cells that may be present in samples before the Fas antibody treatment, we use the annexin-negative or "live cell" population within each subset to assess the amount of cell death with and without treatment (see Note 15).

Table 1
Worktable for calculation of specific cell death from % live cells

(a) % Live cells in treated samples	CD4+/CD45RA−/annexin- : APO 1–3 at 1,000 ng/mL	=47.8 %
(b) % Live cells in untreated samples	CD4+/CD45RA−/annexin- : APO 1–3 at 0 ng/mL	=69.7 %
(c) Specific cell death = 1 − 47.8/69.7 = 0.3142		
% Specific cell death = 31.42 %		

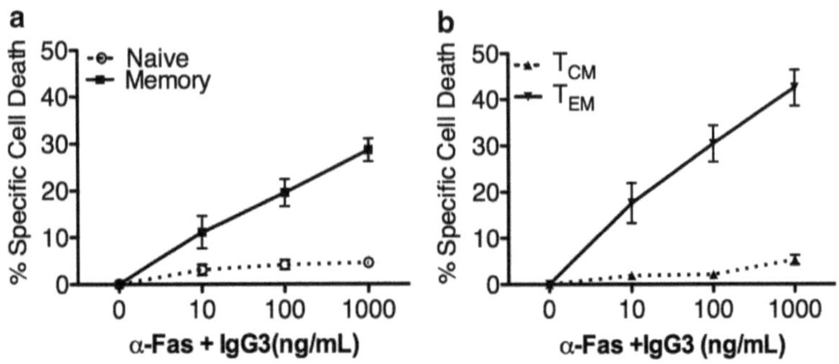

Fig. 3. Fas-induced apoptosis within ex vivo naïve and memory CD4+ T cell subsets using the multiparameter flow analysis. Ex vivo PBMCs from normal donors were treated with anti-Fas APO-1.3 (cross-linked with IgG3) at the indicated dose–response curve for 6–8 h. Cells were then surface stained for CD4+ subset markers along with annexin V and a multiparameter flow cytometry analysis was done. The specific cell death within naïve and memory (**a**) or T_{CM} and T_{EM} memory (**b**) CD4+ T cells was calculated from the % of annexin V− "live" cells within each gated subset by using the formula described in the text.

6. We use the following calculation to assess percentage of "specific cell death" within subsets of CD4+ T cells that have been treated or untreated with anti-Fas antibodies:

$$\text{Specific cell death} = 1 - \frac{\%\text{ live cells in treated sample}}{\%\text{ live cells in untreated sample}}.$$

7. As an example, we have used part of the data set from a representative ex vivo experiment to demonstrate % specific cell death calculation in the memory T cell pool when treated with 1,000 ng/mL of APO-1.3 + IgG3 (Table 1).
8. A dose–response curve of % specific cell death induced by APO-1.3 treatment in ex vivo CD4+ T cell subsets from a mixed population of PBMCs (from normal donors), with typical reactivity to Fas-induced apoptosis within T cell subsets, is shown in Fig. 3 (see Note 16).

4. Notes

1. Whole blood collected with acid citrate dextrose (ACD) as the anticoagulant usually results in better viability of the PBMCs compared to other anticoagulants and therefore will give less background cell death in the subsequent assay.
2. Alternatively, the blood/PBS mix can be layered on top of the Ficoll. Pipet 13 mL of Ficoll into 50 mL conical tube and then slowly pipet blood above the Ficoll layer, being careful not to disrupt the interface.
3. Performing the washing steps promptly and thoroughly is important for removing all Ficoll, since the remaining Ficoll can affect the viability of the cells and thus increase background cell death in the assay.
4. Recombinant Fas ligand (FLAG-tagged FasL, Enzo Life Sciences) can be used in place of the anti-Fas antibody. However, APO-1.3 is usually more effective and reliable. Cross-linking of APO-1.3 is critical to successfully induce cell death. If using Fas ligand, it must also be oligomerized for effective apoptosis induction. Recombinant FLAG-tagged FasL can be cross-linked with anti-FLAG antibody (Sigma-Aldrich).
5. Due to subsequent staining with antibodies for flow cytometry, protein A or protein G is not used as a cross-linker. APO-1.3 is a mouse IgG_3 antibody; thus using anti-IgG_3 allows for specific cross-linking of APO-1.3 and not the antibodies used for flow cytometry.
6. Staurosporine, a protein kinase inhibitor that activates caspases, can be used as a positive control to ensure apoptosis induction for the annexin V staining control used for single-color compensation.
7. Incubation time can be up to 12 h but 6–8 h is optimal. Longer incubation times beyond 6 h will lead to more late-stage apoptosis and secondary necrosis which can result in nonspecific staining and cell loss.
8. From this point on, cells can be processed on the bench in a non-sterile manner. A multichannel pipette will work best for transferring cells.
9. It is important to use only ABB with 0.5 % BSA for all the washes as well as staining until the cells are fixed, since binding of annexin V to phosphatidylserine is a reversible reaction that specifically requires the presence of Ca^{2+} in the buffer.
10. We recommend using the CCR7 FITC clone# FAB197F from R&D for maximal sensitivity in CCR7 staining, since anti-CCR7 is usually a weakly staining antibody. An example of a typical

Table 2
Recommended dilutions for setting up a multiparameter FACS stain

Fluorochrome	Dilution
CCR7 FITC (R&D)	1:50
CD45RA PE	1:200
CD4 PE-Cy5.5	1:200
Annexin APC (Caltag)	1:60
CD27 APC eflour-780	1:200

multiparameter staining mix (single-color controls: it is important to maintain similar cell numbers, antibody dilutions, and buffer volume while staining for single colors to optimize compensations) with recommended dilutions of antibody is indicated in Table 2.

11. At this point, if directly proceeding to FACS analysis of non-fixed cells, a live–dead marker 4′, 6-diamidino-2-phenylindole (DAPI) or propidium iodide (PI) can be added to ABB when resuspending stained cells. This will additionally differentiate cells undergoing active apoptosis (annexin V+/DAPI OR PI−) versus dead/necrotic cells (annexin V^+/DAPI OR PI^+) and is useful when the PBMCs have a low initial viability.

12. Given the number of markers used in the multiparameter assay and to further allow malleability in changing flourophores as well as expanding the assay parameters for other immune cell subsets, it is preferable to analyze stained cells using a 2- or 3-laser flow cytometer capable of analyzing at least 8–10 colors (BD FACSCanto, BD FACSVerse, BD LSR II).

13. Alternatively, the marker CD45RO could substitute for differentiating naïve ($CD4^+CD45RO^-$) and memory ($CD4^+CD45RO^+$) CD4 subsets.

14. In order to normalize the gating to define $CCR7^+/CD27^+$ central memory cells within the memory $CD4^+$ T cells, apply the gates first to the naïve population of cells, which always tend to remain $CCR7^+/CD27^+$. Application of this same gate from the naïve to memory subsets will better differentiate the T_{CM} cells.

15. In unfixed cells where live cell markers like DAPI or PI have additionally been used, the annexin $V^-/DAPI^-$ or PI^- population would be considered "live cells" for calculation of cell death.

16. In experiments where CCR7 staining is minimal or absent, CD27 staining alone can be used to delineate T_{CM} from T_{EM}. Unpublished data from our lab (Ramaswamy et al.) indicates that transitional memory subsets (defined as $CCR7^-/CD27^+$ cells) have a similar sensitivity as the T_{CM} subset to Fas-induced apoptosis.

Acknowledgements

This work was supported by the Intramural Research Program of National Institute of Allergy and Infectious Diseases and National Institute of Arthritis and Musculoskeletal and Skin Diseases.

References

1. Siegel RM, Chan FK, Chun HJ, Lenardo MJ (2000) The multifaceted role of Fas signaling in immune cell homeostasis and autoimmunity. Nat Immunol 1:469–474
2. Fisher GH, Rosenberg FJ, Straus SE, Dale JK, Middleton LA, Lin AY, Strober W, Lenardo MJ, Puck JM (1995) Dominant interfering Fas gene mutations impair apoptosis in a human autoimmune lymphoproliferative syndrome. Cell 81:935–946
3. Straus SE, Sneller M, Lenardo MJ, Puck JM, Strober W (1999) An inherited disorder of lymphocyte apoptosis: the autoimmune lymphoproliferative syndrome. Ann Intern Med 130:591–601
4. Watanabe-Fukunaga R, Brannan CI, Copeland NG, Jenkins NA, Nagata S (1992) Lymphoproliferation disorder in mice explained by defects in Fas antigen that mediates apoptosis. Nature 356:314–317
5. Ramaswamy M, Cruz AC, Cleland SY, Deng M, Price S, Rao VK, Siegel RM (2011) Specific elimination of effector memory CD4+ T cells due to enhanced Fas signaling complex formation and association with lipid raft microdomains. Cell Death Differ 18:712–720
6. Hughes PD, Belz GT, Fortner KA, Budd RC, Strasser A, Bouillet P (2008) Apoptosis regulators Fas and Bim cooperate in shutdown of chronic immune responses and prevention of autoimmunity. Immunity 28:197–205
7. Oliveira JB, Bleesing JJ, Dianzani U, Fleisher TA, Jaffe ES, Lenardo MJ, Rieux-Laucat F, Siegel RM, Su HC, Teachey DT, Rao VK (2010) Revised diagnostic criteria and classification for the autoimmune lymphoproliferative syndrome (ALPS): report from the 2009 NIH International Workshop. Blood 116:e35–e40
8. Turbyville JC, Rao VK (2010) The autoimmune lymphoproliferative syndrome: a rare disorder providing clues about normal tolerance. Autoimmun Rev 9:488–493
9. Kovacs B, Vassilopoulos D, Vogelgesang SA, Tsokos GC (1996) Defective CD3-mediated cell death in activated T cells from patients with systemic lupus erythematosus: role of decreased intracellular TNF-alpha. Clin Immunol Immunopathol 81:293–302
10. Snow AL, Marsh RA, Krummey SM, Roehrs P, Young LR, Zhang K, van Hoff J, Dhar D, Nichols KE, Filipovich AH, Su HC, Bleesing JJ, Lenardo MJ (2009) Restimulation-induced apoptosis of T cells is impaired in patients with X-linked lymphoproliferative disease caused by SAP deficiency. J Clin Invest 119: 2976–2989

Chapter 5

Visualization of Fas-Mediated Death-Inducing Signaling Complex Formation by Immunoprecipitation

Nelia Cordeiro and Nicolas Bidère

Abstract

The prototypical death receptor Fas (also known as CD95 or Apo-1) plays an essential role in the maintenance of lymphocyte homeostasis. Propagation of cell death through Fas relies on the formation of a multiprotein complex at the receptor level known as the death-inducing signaling complex (DISC). Here, we describe an immunoprecipitation-based protocol to study DISC assembly in activated human T lymphocytes. This procedure is a useful tool to visualize proteins associated with Fas.

Key words: Fas/CD95/APO-1, Death-inducing Signaling complex (DISC), Human T lymphocytes, Immunoprecipitation, SDS-PAGE, Semi-dry transfer, Immunoblots

1. Introduction

The TNF receptor superfamily member Fas is a prototypical death receptor that participates in the physiological elimination of activated lymphocytes during an immune response (1). Fas shares with other death receptors an intracellular segment of 80 amino acids termed the death domain, which is essential for transmitting cell death (2). Once engaged by its cognate ligand Fas-L, Fas assembles a death-inducing signaling complex (DISC) at the receptor level (3). First, the adaptor FADD is recruited to the receptor through homotypic interactions between the death domains present in both Fas and FADD. In turn, FADD recruits and activates the cysteinyl protease caspase-8, which unleashes the apoptotic program and leads to the cell's demise (4). Genetic defects affecting Fas signaling lead to the development of an autoimmune lymphoproliferative syndrome (ALPS), illustrating the importance of Fas in the maintenance of immune homeostasis (5).

Besides FADD and caspase-8, numerous proteins, such as FLIP (6), caspase-10 (7), RIP1 (8), or caspase-2 (9), were later identified

Andrew L. Snow and Michael J. Lenardo (eds.), *Immune Homeostasis: Methods and Protocols*, Methods in Molecular Biology, vol. 979, DOI 10.1007/978-1-62703-290-2_5, © Springer Science+Business Media New York 2013

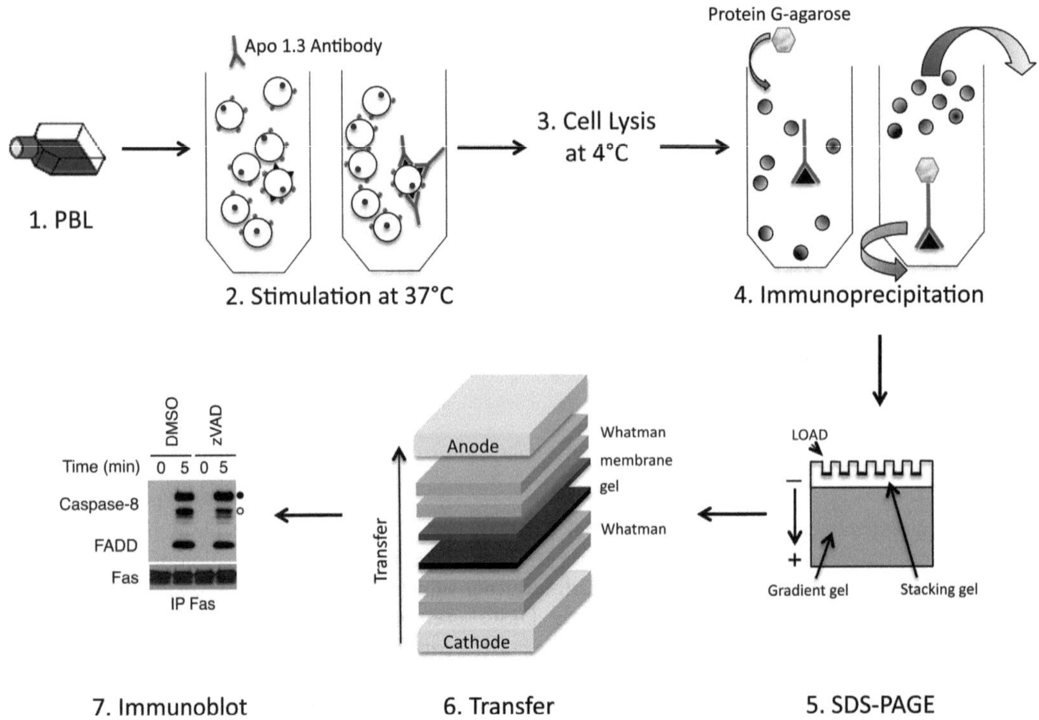

Fig. 1. General overview of DISC immunoprecipitation procedure. Activated human peripheral blood lymphocytes (PBL) are cultured at 37°C (1). Cells are exposed to Fas agonistic antibodies at 37°C (2). Cell extracts are prepared with lysis buffer at 4°C (3) and incubated with Protein G agarose under rotation (4). Immunocomplexes are washed and lysed prior to analysis by SDS-PAGE (5). Proteins are transferred onto nitrocellulose membranes and examined by immunoblot (6, 7). In (7), cells were preincubated with the pan caspases inhibitor zVAD or vehicle DMSO (20 μM, 30 min) prior to Apo1.3 stimulation for 5 min. Solid and open circles indicate full-length and processed caspase-8. IP, immunoprecipitation. Note that caspase inhibition did not affect FADD and caspase-8 recruitment.

as part of the DISC. Here, we provide an easy and reproducible procedure to visualize proteins bound to Fas in lymphocytes (see Fig. 1 for general overview). This protocol uses an agonistic Fas antibody (Apo1.3) that serves to both activate Fas signaling and precipitate the DISC. DISC immunoprecipitation is a powerful tool to study both well-established and yet-to-be-discovered Fas partners.

2. Materials

2.1. Cell Culture

1. Peripheral blood lymphocytes (PBL) from healthy voluntary donors. T lymphocytes may be stimulated as defined in Chapters 2 and 3 (10).
2. Culture medium: RPMI 1640-GlutaMax™-I (Gibco), 1 M HEPES, 100 mM sodium pyruvate, 100× MEM nonessential amino acids, penicillin (10,000 U/mL)/streptomycin (10,000 μg/mL), heat-inactivated fetal calf serum (see Note 1).

For 500 mL of RPMI 1640-GlutaMax™-I, add 5 mL of 1 M HEPES, 5 mL of 100 mM sodium pyruvate, 5 mL of 100× MEM nonessential amino acids, 5 mL of penicillin (10,000 U/mL)/streptomycin (10,000 μg/mL), and 50 mL of heat-inactivated fetal calf serum. Store at 4°C.

3. 1× Dulbecco's phosphate-buffered saline (PBS).

2.2. Antibodies

Anti-Fas antibody, clone Apo1.3 (mouse IgG3), can be purchased from Axxora (ALX-805-020-C100). Antibodies against Fas, FADD, and caspase-8, as well as secondary antibodies conjugated to horseradish peroxidase (HRP), can be obtained from numerous companies (see Note 2).

2.3. Immunoprecipitation Components

1. 1 M Tris–HCl pH 7.5: Weight 121.1 g Tris (hydromethyl) aminomethane, and transfer to a glass beaker. Add ultrapure water to a volume of about 900 mL. Mix and adjust pH with HCl. Make up to 1 L with ultrapure water into a graduated cylinder. Store at room temperature.
2. NaCl 5 M: Weight 292.2 g NaCl and transfer to a graduated cylinder. Make up to 1 L with ultrapure water. Mix and store at room temperature.
3. Lysis buffer stock: 30 mM Tris/HCl pH 7.5, 150 mM NaCl, 1% Triton X-100, and 10% glycerol. For 500 mL: 15 mL Tris/HCl 1 M pH 7.5, 15 mL NaCl 5 M, 5 mL of Triton X-100, 50 mL of glycerol, and ultrapure water up to 500 mL. Lysis buffer is further filtered, and stored at room temperature.
4. Complete cocktail inhibitors tablet (Roche). Dissolve a tablet of complete proteases inhibitors into 50 mL of prepared lysis buffer before use. Store at 4°C (see Note 3).
5. Protein G agarose beads (see Note 4).

2.4. Immunoblotting Components

1. SDS-PAGE lysis buffer: 450 mM Tris–HCl pH 8.45, 12% glycerol, 4% SDS, 0.0075% Coomassie blue G, 0.0025% phenol red, 10% β-mercaptoethanol (see Note 5).
2. Molecular weight protein marker: Precision Plus Protein All Blue Standards (Bio-Rad) or equivalent.
3. Running buffer: 25 mM Tris–HCl pH 8.3, 192 mM glycine, 0.1% SDS. For 2 L, weight 6.055 g of Tris, and 28.8 g of glycine. Dissolve with ultrapure water and add 2 mL of 10% SDS. Mix and store at room temperature.
4. Precast polyacrylamide gels: NuPAGE® Novex® 4–12% Bis-Tris Gels (1.0 mm thick, 10-well) from Invitrogen.
5. Transfer buffer: 12 mM Tris–HCl pH 8.3, 96 mM glycine, and 20% ethanol. For 2 L, weight 2.9 g of Tris and 14.4 g of glycine. Dissolve with ultrapure water up to 1.6 L. Add 400 mL of ethanol. Mix and store at room temperature (see Note 6).

6. Semi-dry transfer apparatus (Thermo Scientific Owl Hep-1 or equivalent).
7. Whatman paper.
8. Nitrocellulose membranes (Hybond ECL, Amersham).
9. Ponceau S solution (Sigma).
10. Washing solution: PBS containing 0.1% Tween-20 (PBS-T). For 2 L, add 200 mL of PBS 10× into a 2 L graduated cylinder. Complete up to 2 L with ultrapure water. Add 2 mL of Tween-20. Mix and store at room temperature.
11. Blocking and diluent solution: 5% skimmed milk in PBS-T. For 100 mL, weight 5 g of skimmed milk and transfer to a 100 mL graduated cylinder. Make up to 100 mL with PBS-T. Store at 4°C.
12. Chemiluminescent HRP substrate: Immobilon (Millipore).
13. Autoradiography film and film developing equipment.

3. Methods

3.1. Cell Culture, Lysis Buffer Preparation, and Stimulation

1. Enumerate cells, and aliquot 1×10^7–2.5×10^7 cells in 5 mL culture media per condition.
2. Stimulation is achieved with 1 μg/mL anti-Apo1.3 at 37°C (see Note 7).

3.2. Cell Lysis

Carry out all procedures at 4°C unless otherwise specified.

1. Immediately following stimulation, dilute stimuli by adding ice-cold PBS. Centrifuge for 5 min at $500 \times g$ at 4°C. Discard the supernatants and transfer cells into 1.5 mL microfuge tubes with 0.5–1 mL 1× PBS.
2. Centrifuge as above, and eliminate supernatants.
3. Lyse cellular pellets in 1 mL of lysis buffer on ice for 15 min.
4. Centrifuge for 15 min at $10{,}000 \times g$, and collect supernatants containing cell extracts in new microfuge tubes (see Note 8).

3.3. Immunoprecipitation

Perform all steps at 4°C unless otherwise specified (see Note 9).

1. Protein G agarose preparation: 20–50 μL Protein G agarose per immunoprecipitation are washed in 500 μL lysis buffer 4 times by centrifugation at $10{,}000 \times g$ for 10 s (see Note 10).
2. Mix cell extract and washed Protein G agarose in microfuge tubes.
3. Add 2 μg Apo1.3 to the control untreated lysates (see Note 11).

4. Place under rotating agitation for at least 1 h at 4°C (see Note 12).
5. Centrifuge for 10 s at 10,000 ×*g* (see Note 13).
6. Discard the supernatant, add 0.5–1 mL of lysis buffer, and centrifuge at 10,000 ×*g* for 10 s.
7. Repeat wash in step 6 three additional times, and eliminate the supernatant.
8. Add 30 μL of Tris–Glycine SDS Sample Buffer (with β-mercaptoethanol) directly on Protein G agarose-immune complex pellets (see Note 14).
9. Heat all samples at 95°C for 5 min. Centrifuge for 30 s at 10,000 ×*g* (see Note 15).

3.4. SDS-PAGE and Transfer onto Nitrocellulose Membranes

1. Load samples as well as protein standards on the gel.
2. Electrophorese at 140 V until the sample has entered the stacking gel and switch to 170 V until the dye front leaks out from the bottom of the gel.
3. Following electrophoresis, open the gel plates with a spatula, and discard the stacking gel as well as the front migration. Place the gel into a clean container filled with transfer buffer. Cut a membrane and 2 pieces of Whatman paper to the size of the gel and soak with transfer buffer. Assemble a Whatman–gel–membrane–Whatman sandwich. With a semi-dry transfer apparatus, transfer at room temperature for 1 h at 300 mA per gel (see Note 16).

3.5. Immunoblotting

1. Right after transfer, place the membrane in a tray containing Ponceau S solution for 5 min with gentle agitation (see Note 17).
2. Block the membranes by incubating for 1 h with blocking solution under agitation at room temperature.
3. Incubate membranes with appropriate primary antibodies for at least 1 h under agitation at room temperature (see Note 18).
4. Wash membranes 4 times for 5 min with PBS-T under agitation.
5. Add appropriate secondary antibodies conjugated to HRP in diluent solution for 45 min under agitation at room temperature (see Note 18).
6. Wash membranes 4 times for 5 min with PBS-T under agitation.
7. Remove excess of PBS-T, and incubate the membranes with chemiluminescent HRP substrate for 5 min at room temperature.
8. Detection of proteins of interest is achieved with X-ray film or with a chemiluminescence-sensitive camera.

4. Notes

1. To inactivate complement, heat fetal calf serum at 56°C for 30 min. Serum is filtered prior to use.
2. We used anti-FADD and anti-caspase-8 sold by BD Biosciences, and anti-Fas was from Santa Cruz Biotechnology. Secondary antibodies coupled to HRP can be purchased from Southern Biotech.
3. We found that buffer containing protease inhibitors can be stored at 4°C for several days.
4. Protein A agarose, which also precipitates mouse IgG3, can also be used.
5. We purchase a ready-to-use Tris–Glycine SDS Sample Buffer 2× from Invitrogen, and add fresh 2β-mercaptoethanol (Sigma) to get a 10% final concentration.
6. We found it more convenient to buy a 25× stock of Tris–Glycine Transfer Buffer. For 1 L, transfer 40 mL of 25× stock into a 1 L graduated cylinder containing 200 mL of pure ethanol. Add 760 mL of ultrapure water. Keep at room temperature.
7. The DISC assembles within minutes following Fas ligation. We therefore recommend kinetic experiments ranging from 5- to 30-min stimulation.
8. Identical volume aliquots, typically 10 μL, can be saved at this stage. These constitute the "input" lysates.
9. More details regarding immunoprecipitation can be found in ref. (11).
10. We find it easier to collect the total volume of Protein G agarose in one single 1.5 mL microfuge tube, and wash it as a bulk. During the final wash, Protein G agarose is split into tubes and centrifuged. Washing Protein G agarose is essential to remove the ethanol contained in the stock.
11. This step is essential to immunoprecipitate Fas from untreated samples.
12. Performing this immunoprecipitation step overnight at 4°C is also possible.
13. Prior to discarding the supernatants, aliquots can be saved. These constitute "output" lysates. Immunoblotting Fas in both input and output lysates allows you to gauge the immunoprecipitation efficiency.
14. Mix 10 μL of Tris–Glycine SDS Sample Buffer with β-mercaptoethanol with input and output extracts.
15. At this step, samples can be stored at −20°C for future use, or further resolved by SDS-PAGE.

16. We used a semi-dry transfer apparatus (Thermo Scientific Owl Hep-1). To eliminate air bubbles, we roll a 5 mL plastic pipette between each step of the sandwich.
17. This step allows you to visualize proteins and gauge transfer efficiency from gel to membrane. Note that Apo1.3 heavy and light chains should be seen in the immunoprecipitation lanes. Ponceau S is easily washed away with ultrapure water or 1× PBS.
18. Primary and secondary antibodies were classically used at a 1:1,000 and 1:5,000 dilution in diluent solution, respectively. In addition, primary antibodies can also be left overnight at 4°C under agitation. Because FADD (23 kDa) and caspase-8 (55 kDa) size are significantly different, primary antibodies can be incubated simultaneously.

Acknowledgements

This work was supported by grants from ANR JCJC, Fondation de France, Association pour la Recherche contre le Cancer, and Ligue Nationale contre le Cancer.

References

1. Krammer PH, Arnold R, Lavrik IN (2007) Life and death in peripheral T cells. Nat Rev Immunol 7(7):532–542
2. Tibbetts MD, Zheng L, Lenardo MJ (2003) The death effector domain protein family: regulators of cellular homeostasis. Nat Immunol 4(5):404–409
3. Kischkel FC et al (1995) Cytotoxicity-dependent APO-1 (Fas/CD95)-associated proteins form a death-inducing signaling complex (DISC) with the receptor. EMBO J 14(22):5579–5588
4. Peter ME, Krammer PH (2003) The CD95(APO-1/Fas) DISC and beyond. Cell Death Differ 10(1):26–35
5. Bidere N, Su HC, Lenardo MJ (2006) Genetic disorders of programmed cell death in the immune system. Annu Rev Immunol 24: 321–352
6. Thome M et al (1997) Viral FLICE-inhibitory proteins (FLIPs) prevent apoptosis induced by death receptors. Nature 386(6624):517–521
7. Wang J, Chun HJ, Wong W, Spencer DM, Lenardo MJ (2001) Caspase-10 is an initiator caspase in death receptor signaling. Proc Natl Acad Sci U S A 98(24):13884–13888
8. Stanger BZ, Leder P, Lee TH, Kim E, Seed B (1995) RIP: a novel protein containing a death domain that interacts with Fas/APO-1 (CD95) in yeast and causes cell death. Cell 81(4): 513–523
9. Lavrik IN, Golks A, Baumann S, Krammer PH (2006) Caspase-2 is activated at the CD95 death-inducing signaling complex in the course of CD95-induced apoptosis. Blood 108(2): 559–565
10. Snow AL et al (2009) Restimulation-induced apoptosis of T cells is impaired in patients with X-linked lymphoproliferative disease caused by SAP deficiency. J Clin Invest 119(10): 2976–2989
11. Bonifacino JS, Dell'Angelica EC, Springer TA (2001) Immunoprecipitation. Curr Protoc Mol Biol Chapter 10:Unit 10 16.

Chapter 6

Analyses of Programmed Cell Death in Dendritic Cells

Min Chen, Lily Huang, and Jin Wang

Abstract

Dendritic cells (DCs) are professional antigen-presenting cells that can regulate both innate and adaptive immune responses. Programmed cell death of DCs plays an important role in maintaining the homeostasis of DCs and in the regulation of immune responses. Methods to measure the rate of spontaneous and T cell-mediated cell death of DCs are described here. These procedures can be adapted to study the induction of dendritic cell death in different settings of immune responses.

Key words: Antigen presenting cell, Dendritic cell, Myeloid, Plasmacytoid, Spontaneous cell death, T-cell induced cell death, Ovalbumin, BrdU

1. Introduction

After differentiation, DCs lose their proliferation potential and undergo high rates of cell death in vivo. Programmed cell death of DCs plays an important role in the maintenance of DC homeostasis. Defective cell death of DCs leads to abnormal DC accumulation, uncontrolled activation of lymphocytes, and disruption of immune tolerance (1, 2). We summarize the procedures for determining the rates of cell death of DCs in mice by different in vitro and in vivo methods, including measuring spontaneous and T cell-induced cell death of DCs, estimating the rates of clearance of DCs in vivo in the presence or the absence of antigen-specific T cells by adoptive transfer, as well as determining the half-life of resident DCs in lymphoid organs by BrdU labeling (3–5).

Andrew L. Snow and Michael J. Lenardo (eds.), *Immune Homeostasis: Methods and Protocols*, Methods in Molecular Biology, vol. 979, DOI 10.1007/978-1-62703-290-2_6, © Springer Science+Business Media New York 2013

2. Materials

1. Liberase TL research grade (Roche): Dissolve 5 mg in 2 ml ddH_2O (double-distilled or Milli-Q), aliquot, and store at −20 °C. This solution is stable for 2–3 months. Dilute to 0.32 mg/ml in serum-free medium prior to use.
2. Streptavidin-conjugated BioMag beads (Qiagen).
3. MACS beads conjugated to anti-PDCA-1, anti-CD11c, anti-CD8, or anti-CD4 (Miltenyi Biotec).
4. Biotinylated antibody against CD3, CD4, CD8, I-A^b, B220, Thy1.2, CD19, IgM, CD49b, or TER119 (BD Biosciences).
5. Magnetic stand (BD Biosciences).
6. LS columns (Miltenyi Biotec).
7. 40 μm strainer and 5 ml syringes.
8. ACK lysis buffer: 0.15 M NH_4Cl, 1 mM $KHCO_3$, 0.1 mM EDTA, pH 7.2.
9. Phosphate-buffered saline (PBS)–EDTA buffer: PBS containing 2% FCS and 2 mM EDTA.
10. Fc receptor blocker: 1 μg/ml anti-CD16/CD32 (BD Biosciences) plus 10 μg/ml rat IgG (Sigma).
11. 25-gauge needles.
12. Cytokines: Recombinant mouse GM-CSF, mouse IL-4 (Invitrogen).
13. DC medium: RPMI 1640 medium containing 10% fetal bovine serum (FBS), 10 ng/ml of mouse GM-CSF and IL-4, and antibiotics (penicillin–streptomycin).
14. Lympholyte-M cell separation medium (Ficoll, Accurate Chemical).
15. Mice: OT1 transgenic mice (Jackson Laboratory) carrying TCR transgenes specific for ovalbumin (OVA) peptide $OVA_{257–264}$ "SIINFEKL" presented by H-2K^b; OT2 transgenic mice (Jackson Laboratory) carrying TCR specific for $OVA_{323–339}$ peptide "ISQAVHAAHAEINEAGR" presented by I-A^b. Unless specified otherwise, 5–8-week-old C57BL6 mice can be used for DC preparation (Jackson Laboratory).
16. OVA peptides: $OVA_{257–264}$ peptide and $OVA_{323–339}$ peptide.
17. Carboxyfluorescein succinimidyl ester (CFSE, Invitrogen): Dissolve CFSE in DMSO to make 5 mM stock, aliquot, and freeze at −20 °C.
18. 7-Amino-Actinomycin D (7-AAD) staining solution (BD Biosciences).
19. 10× Annexin V Binding Buffer: 0.1 M HEPES, pH 7.4; 1.4 M NaCl; 25 mM $CaCl_2$. Dilute to 1× prior to use.

20. FITC-Annexin V (BD Biosciences).
21. Propidium iodide (PI) solution (200 μg/ml): Dissolve 10 mg PI (Sigma P1470-25 mg) in 50 ml ddH_2O.
22. BrdU (Sigma): Dissolve in PBS to obtain 10 mg/ml or 0.8 mg/ml solution.
23. FITC-anti-BrdU Flow kit (BD Biosciences).
24. Flow cytometry staining buffer: PBS with 2% FCS and 0.01% sodium azide.
25. Blocking buffer: Staining buffer +10 μg/ml Rat IgG +1 μg/ml anti-CD16/32.
26. PE-conjugated anti-PDCA-1 (Miltenyi Biotec).
27. Antibody cocktails: PE–anti-CD11c, PE–Cy5-anti-CD8, and antigen-presenting cell (APC)–anti-CD4; APC–anti-CD11c and PE–anti-CD11b (BD Biosciences).
28. Flow cytometer, e.g., LSRII or comparable instrument.

3. Methods

Unless otherwise noted, all centrifugation steps are performed at ~400 × *g* for 5 min.

3.1. Purification of Spleen DC Subsets from Mice

1. Liberase treatment: Inject 1 ml (0.32 mg/ml) Liberase per mouse spleen. Then, cut the spleens into small pieces. Incubate at room temperature for 10–15 min.
2. Make a single-cell suspension by using a cell strainer and pressing the spleen pieces through with a 5 ml syringe plunger. Add 10× volume of PBS containing 2% FCS to rinse the cells through the strainer. Harvest cells into a tube and pellet the cells by centrifugation (~400 × *g*).
3. To lyse red blood cells, resuspend the cell pellet in ACK lysis buffer and leave the cells at room temperature (RT) for 2 min. Stop the lysis by adding 10× volume of PBS containing 2% FCS and pellet the cells by centrifugation.
4. Wash the cells with PBS containing 2% FCS. Pass the cells through a 40 μm strainer. Take an aliquot for cell counting. Pellet the cells by centrifugation.
5. Resuspend the cell pellet in PBS–EDTA buffer (10^8 cells/ml).
6. To prevent nonspecific binding of antibodies to Fc receptors, incubate the cells with anti-CD16/32 (1 μg/ml) and rat IgG (10 μg/ml) (Fc receptor blocker) on ice for 10 min.
7. Add 1 μg/ml biotinylated antibodies against CD3, Thy1.2, CD19, IgM, CD49b, and TER119 and incubate the cells on ice for 30 min.

8. During the antibody incubation, wash Streptavidin-BioMag beads 3 times with an appropriate volume of PBS–EDTA buffer. Use 750 μl beads/10^9 cells in 8 ml PBS–EDTA buffer.
9. Wash cells once by adding 20× volume of PBS–EDTA buffer, and centrifuge for 5 min at $400 \times g$.
10. Resuspend the antibody-labeled cells in PBS–EDTA buffer (10^8 cells/ml). Add the required amount of Streptavidin-BioMag beads (see step 8). Incubate at 4–8 °C for 30 min. Invert the tube every 10 min to mix.
11. Remove cells bound to the magnetic beads using the magnetic stand (see Note 1). Transfer the cells in the supernatant (unbound to the magnetic beads) to a new tube. Repeat the same step 3 times. Pellet the unbound fraction of cells (i.e., enriched DCs) by centrifugation (~$400 \times g$).
12. Equilibrate a MACS LS column with 3 ml PBS–EDTA buffer. Resuspend the cells in 3 ml PBS–EDTA buffer and pass the cells through the LS column to remove any residual magnetic beads. Wash column once with 3 ml PBS–EDTA buffer. Collect the flow-through fraction and pellet those cells by centrifugation.
13. To isolate PDCA-1^+ plasmacytoid DCs (pDCs), resuspend the cells (10^8) in 400 μl PBS–EDTA buffer. Incubate with Fc blocker and 50 μl anti-PDCA-1 MACS beads (for up to 20 spleens) at 6–12 °C for 15 min. Wash cells with 20× volume of buffer. Centrifuge for 5 min.
14. Resuspend the cells in 3 ml PBS–EDTA buffer. Pass the cells through the pre-equilibrated LS column. Wash the column 3× with 3 ml PBS–EDTA buffer. Collect the flow-through fraction and the first 2 washes in a separate tube for the next step. After removing the column from the magnet, elute the pDCs from the column in 5 ml PBS–EDTA buffer into a separate tube. Pellet the pDCs and the flow-through fraction by centrifugation ($400 \times g$).
15. To isolate CD11c^+ DCs, resuspend cells in the flow-through fraction in 400 μl PBS–EDTA buffer and incubate with the Fc receptor blocker and 100 μl anti-CD11c MACS beads (for up to 20 spleens) at 6–12 °C for 15 min. Wash cells with 20× volume of buffer.
16. Equilibrate a MACS LS column with 3 ml PBS–EDTA buffer. Resuspend the cells in 3 ml buffer. Pass the cells through the LS column. Wash column 3× with 3 ml PBS–EDTA buffer. Discard the flow-through fraction. Remove the column from the magnet and elute CD11c^+ DCs from the column with 5 ml PBS–EDTA buffer. Pellet DCs by centrifugation.
17. Resuspend each DC subset in an appropriate volume of complete RPMI medium and count the cells.

3.2. Generation of Bone Marrow-Derived DCs

1. Collect two femurs and two tibias from each mouse (aged 5–8 weeks).
2. Clean the bones with sterile dissecting scissors to remove as much muscle tissue as possible. Place all bones from the same mouse in 15 ml tube with sterile PBS and keep on ice.
3. Move to the tissue culture hood. Aspirate PBS from the tube and immerse the intact bones in 70% ethanol for 2 min to sterilize them. Then wash 3× with PBS in the 15 ml tube.
4. After the last PBS wash, add 10 ml RPMI media and pour all the bones and media out in a 60 mm dish.
5. Clip the head of the bone from shaft, saving the shaft in one dish with RPMI and the heads on the dish cover. Harvest the bone marrow from the shafts by flushing with a 25-gauge needle and 10 ml syringe. Break up the bone marrow harvested from shafts by passing through a 19-gauge needle. Harvest the bone marrow from the bone heads by crushing with a syringe plunger. Transfer all the cells to a clean 15 ml tube.
6. Pellet cells and deplete the red blood cells by resuspending in ACK lysis buffer (0.5 ml for cells from each mouse) for 2 min at room temperature. Stop the lysis by adding 10× volume of PBS containing 2% FCS. Pellet cells by centrifugation.
7. Resuspend the cells in PBS containing 2% FCS. Pass the cells through a cell strainer into a 50 ml conical polypropylene tube to create a single-cell suspension. Count the cells and then pellet them by centrifugation. A typical yield is around $10–30 \times 10^6$ cells per mouse.
8. Blocking of Fc receptors: Resuspend the cells at 1×10^8/ml in PBS–EDTA buffer. Incubate with Fc receptor blocker on ice for 10 min.
9. Antibody incubation: Incubate the cells with biotinylated antibodies (1 μg/ml) against CD4, CD8, Thy1.2 (for removing T cells), biotinylated anti-I-A^b (to remove class II MHC^+ cells, e.g., B cells and macrophages), and biotinylated CD45R/B220 (to remove B cells) on ice for 30 min. Flick the tubes with your finger every 15 min to mix the cells.
10. Add 20× volume of PBS–EDTA buffer to the cell suspension and then pellet the cells by centrifugation. Discard the supernatant.
11. Wash Streptavidin-BioMag beads 3× in PBS–EDTA at the same time (see Subheading 3.1, step 8). Use 75 μl beads/10^8 cells.
12. Resuspend the cells in PBS–EDTA buffer (10^8 cells/ml). Add the Streptavidin-BioMag beads to the cells and incubate at 4–8 °C for 30 min. Invert the tube every 10 min to mix.

13. Remove the cells bound to magnetic beads using the magnetic stand. Transfer cells in the supernatant (unbound to the magnetic beads) to a new tube. Repeat the same step 3 times, and then pass the cells through the strainer the last time. Pellet the cells by centrifugation.
14. Resuspend the cells at 3×10^6/ml in DC media to spur differentiation of bone marrow-derived DCs. Culture the cells in 6-well plates (3 ml/well) or in 24-well plates (1.5 ml/well).
15. Feed the cells every two or three days by aspirating half of the media off the top of each well without disturbing the cells at the bottom, and replace with fresh DC media.
16. The resulting DCs should be used between days 6 and 8 (see Notes 2–3).

3.3. T Cell-Mediated Lysis of DCs

3.3.1. Activation of Antigen-Specific CD4+ T Cells (see Notes 4, 5)

1. Add 5 ml PBS containing 2% FCS into one well of 6-well plate. Collect spleens and lymph nodes (LNs) from each OT2 mouse using dissecting scissors and forceps. Make single-cell suspension by using a cell strainer and pressing with a 5 ml syringe plunger. Harvest cells into a 15 ml tube. Pellet the cells by centrifugation.
2. Resuspend cells in ACK lysis buffer to lyse red blood cells and leave at RT for 2 min. Stop the lysis reaction by adding 10 times volume of PBS containing 2% FCS and pellet the cells by centrifugation.
3. Discard the supernatant. Wash the cells with PBS containing 2% FCS. Pass cells through a 40 μm strainer. Take an aliquot for counting. Pellet the cells by centrifugation.
4. Resuspend cell pellet at 2×10^8 cells/ml in PBS–EDTA buffer. Incubate with Fc receptor blocker on ice for 10 min.
5. Incubate with 50 μl anti-CD4 MACS beads/10^8 cells at 6–12 °C for 15 min. Wash cells once with 20× volume of PBS–EDTA buffer. Centrifuge for 5 min.
6. Equilibrate a MACS LS column with 3 ml PBS–EDTA buffer. Resuspend the cell pellets in 3 ml PBS–EDTA buffer and pass cells through the LS column. Wash the column 3× with 3 ml PBS–EDTA buffer. Discard the flow-through fraction. Remove the column from the magnet and elute the column-bound $CD4^+$ cells using 5 ml PBS–EDTA buffer. Pellet cells by centrifugation.
7. Resuspend $CD4^+$ cells in RPMI complete medium. Count the cells.
8. Stimulate $CD4^+$ OT2 T cells (10^6/ml) with DCs (10^5/ml) pulsed with $OVA_{323–339}$ peptide (see Subheading 3.3.3) for 4 days.

3.3.2. Activation of Antigen-Specific $CD8^+$ T Cells (see Notes 4, 5)

1. Purify $CD8^+$ cells from OT1 mice using the same procedures as those used for $CD4^+$ cell purification (Subheading 3.3.1), except that anti-CD8 MACS beads will be used in step 5.
2. Activate $CD8^+$ OT1 T cells (10^6/ml) by stimulating with DCs (10^5/ml) pulsed with $OVA_{SIINFEKL}$ for 4 days.

3.3.3. Pulse DCs with Antigenic Peptide (see Notes 4, 5)

1. Harvest day 7 culture of bone marrow-derived DCs (see Subheading 3.2). Pipet the culture up and down gently to collect loosely attached cells. Wash the plate once with 5 ml PBS. Combine the cells together and centrifuge for 5 min.
2. Ficoll-gradient separation: Resuspend cell pellet in 5 ml PBS–EDTA buffer in a 15 ml tube. Using a pipet, add 2 ml Ficoll to the bottom of the tube slowly without disturbing the cell suspension. Centrifuge at RT at ~700 × g for 15 min. Transfer the live cells at the interphase (top of Ficoll layer) to a new 15 ml tube. Fill the tube with PBS to 15 ml and centrifuge for 5 min.
3. Wash cells once with 5 ml PBS–EDTA buffer and centrifuge for 5 min.
4. Blocking: Resuspend the cells (up to 10^8) in 450 μl buffer plus Fc receptor blocker for 10 min on ice.
5. MACS beads binding: Add 50 μl anti-CD11c MACS beads, and incubate at 6–12 °C for 15 min. Wash cells once with 20× volume of PBS–EDTA buffer. Centrifuge for 5 min and resuspend cell pellet in 3 ml buffer.
6. Equilibrate a MACS LS column with 3 ml PBS–EDTA buffer. Pass cells through the column. Wash column 3 times with 3 ml PBS–EDTA buffer each. Discard the flow-through fraction. Remove the column from the magnet and elute $CD11c^+$ cells from the column using 5 ml buffer and pellet the cells by centrifugation.
7. Resuspend the cells in the DC medium. Count the cells and adjust DCs to 2×10^6 cells/ml.
8. Incubate DCs with 20 μg/ml OVA peptide at 37 °C for 2–3 h. Mix gently every 30 min.
9. Pellet DCs by centrifugation. Wash cells once with 20× volume of PBS–EDTA buffer. Centrifuge again.
10. Resuspend the cells in 3 ml RPMI complete media. Count the cells and adjust DCs to 2×10^6 cells/ml for mixing with OT1 and/or OT2 T cells (see Subheading 3.3.2, step 8).

3.3.4. Harvesting Activated Antigen-Specific T Cells

1. Harvest mixtures of T cells and DCs into a 15 ml tube and centrifuge for 5 min. Discard the supernatant and resuspend the cell pellet in 5 ml PBS–EDTA buffer.

2. Add 2 ml Ficoll to the bottom of the tube slowly without disturbing the cell suspension. Centrifuge at RT, ~700×*g* for 15 min. Transfer live cells at the buffer:Ficoll interface to a new 15 ml tube. Fill the tube with PBS to 15 ml and centrifuge for 5 min.
3. Wash cells once with 5 ml PBS–EDTA buffer and centrifuge for 5 min.
4. Resuspend the cell pellet in 450 μl PBS–EDTA buffer (10^8 cells). Add 50 μl anti-CD11c MACS beads and incubate at 6–12 °C for 15 min. Wash cells once with 20× volume of PBS-EDTA buffer. Centrifuge for 5 min and resuspend the cell pellet in 3 ml buffer.
5. Equilibrate a MACS LS column with 3 ml PBS–EDTA buffer. Pass the cells through the column. Save the flow-through fraction that contains the activated T cells (i.e., $CD11c^-$). Wash the column 3× with 3 ml PBS–EDTA buffer each. Combine the washes with the flow-through fraction and centrifuge for 5 min.
6. Resuspend the cell pellet in RPMI complete medium. Count cells and adjust to 2×10^6/ml.

3.3.5. Set Up the T Cell-Mediated DC Kill Assay

1. Take an aliquot of antigen-pulsed DCs and pellet the cells by centrifugation. Wash cell pellet once with 5 ml PBS.
2. Resuspend the DC pellet in PBS at 2×10^6/ml. Add CFSE to final concentration of 2 μM and incubate at 37 °C for 10 min.
3. Stop the CFSE labeling by adding an equal volume of prewarmed fetal calf serum and incubate cells at 37 °C for 10 min for efflux.
4. Centrifuge for 5 min at 200–300×*g* and discard the supernatant. Wash cells 3 times with PBS containing 2% FCS.
5. Resuspend CFSE-labeled DCs at 2×10^5/ml in complete RPMI medium.
6. Aliquot 100 μl (2×10^4) of DCs per well to 96-well U-bottom plates.
7. Adjust activated T cells to 2×10^6/ml, 6×10^5/ml, and 2×10^5/ml (see Subheading 3.3.4).
8. Set up the kill assay with activated T cells and bone marrow-derived DCs at T:DC ratios of 0:1, 1:1, 3:1, or 10:1—dilute T cells accordingly to achieve the correct ratios. The total volume in each well should be 200 μl. Centrifuge for 5 min.
9. Incubate at 37 °C for 5 h.
10. Collect cells into 5 ml Falcon tubes (Becton Dickinson).
11. Add 5 μl (0.25 μg) 7-(aminoactinomycin D) AAD per tube and incubate for 10 min at RT. Collect cells at a constant flow rate on a flow cytometer for 30 s per sample (Fig. 1).

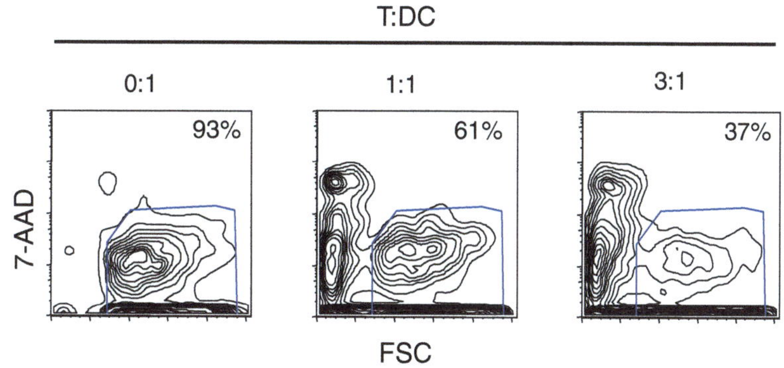

Fig. 1. T cell-dependent killing of DCs. CFSE-labeled DCs were incubated with antigen-specific T cells at different ratios, followed by flow cytometry. $CFSE^{+}$ DCs were gated. Forward scatter (FSC) versus 7-AAD staining plots reveal the loss of viable DCs (FSC^{hi}, $7\text{-}AAD^{lo}$).

12. Percentage of cell loss will be calculated as follows: Percentages of killing of DCs by T cells = 100% × $(DC_{control} - DC_{T})/DC_{control}$, with $DC_{control}$ and DC_{T} representing $CFSE^{+}7\text{-}AAD^{-}$ DCs in the absence or the presence of T cells, respectively.

3.4. Clearance of DCs After Adoptive Transfer

1. Inject purified $CD8^{+}$ OT1 T cells into mice retro-orbitally (2×10^{6}/mouse). Plan to inject enough mice so that there are at least 3 mice per group per time point. Alternatively, you may inject purified $CD4^{+}$ OT2 T cells into mice retro-orbitally (2×10^{6}/mouse) to analyze the effect of $CD4^{+}$ T cells on DC survival.
2. One day later, prepare unpulsed and OVA-pulsed DCs (see Subheading 3.3.3).
3. Resuspend DCs in PBS (2×10^{6}/ml). Add CFSE to final concentration of 5 μM and incubate at 37 °C for 10 min.
4. Stop CFSE labeling by adding an equal volume of pre-warmed fetal calf serum and incubate cells at 37 °C for 10 min for efflux.
5. Centrifuge for 5 min and discard the supernatant. Wash the cells 3× with PBS.
6. Resuspend CFSE-labeled DCs at 2×10^{7}/ml in PBS. Inject mice with 50 μl (1×10^{6}) DCs into the footpad.
7. Harvest the draining popliteal lymph node on days 1, 2, and 5 after DC injection. Treat the lymph node with 0.32 mg/ml Liberase at RT for 10 min.
8. Make a single-cell suspension by using a cell strainer and pressing the node through it with a 5 ml syringe plunger. Add 10× volume of PBS containing 2% FCS. Pass the cells through the same strainer again. Count the cells.

9. Aliquot 2–5 × 10^6 cells into 5 ml tubes and pellet by centrifugation. Resuspend the pellet in 50 μl of Fc blocking buffer and incubate on ice for 10 min.
10. Prepare the antibody cocktail specific for PE-conjugated anti-CD11c (1:100, see Subheading 2) in Fc blocking buffer. Add 50 μl antibody cocktail per tube and mix well. Incubate cells with antibodies on ice for 20 min.
11. Wash cells 1× by adding 2 ml of staining buffer (PBS containing 2% FCS and 1 mM EDTA) per tube. Centrifuge for 5 min at 200–300 × *g* and discard the supernatant.
12. Resuspend the cells in 0.5 ml staining buffer per tube and analyze $CD11c^{+}CFSE^{+}$ cells on a flow cytometer. Quantify the total number of $CD11c^{+}CFSE^{+}$ DCs in the draining lymph node for each mouse.

3.5. Spontaneous Cell Death in DCs

1. Resuspend your selected pool of DCs (see Subheadings 3.1 or 3.2) at 5 × 10^5/ml in RPMI complete medium. Add 100 μl DCs (5 × 10^4/well) to 96-well round-bottom tissue culture plates.
2. Twenty hours after culture, harvest cells to a 5 ml tube. Wash cells 1× by adding 1 ml of PBS per tube, centrifuge for 5 min at 200–300 × *g*, and discard supernatant.
3. Wash the cells 1× with 1 ml of 1× Annexin V Binding Buffer, centrifuge for 5 min, and decant the supernatant by quickly inverting the tube and patting on the paper towel.
4. Resuspend the cell pellets in 30 μl of 1× Annexin V Binding Buffer. Add 2 μl FITC-anti-Annexin V per sample, gently mix the cells, and incubate the cells in the dark for 10 min at RT.
5. Add 1–2 μl PI solution (200 μg/ml) and incubate at RT for another 5 min. The optimal amount of PI may vary depending on experimental systems.
6. Add 400 μl of 1× Annexin Binding Buffer to each tube. Analyze by flow cytometry as soon as possible (within 1 h). Gate on Annexin V^{-} live cells and calculate the percentage loss of live cells as outlined in Subheading 3.3.5 (step 12).

3.6. Measurement of DC Turnover In Vivo (see Notes 6, 7)

1. Inject mice intraperitoneally with 100 μl of 10 mg/ml BrdU solution per mouse. Add BrdU (0.8 mg/ml) to their drinking water for the duration of the experiment.
2. At the desired time point, sacrifice the labeled mice. Isolate the spleen and inject 1 ml Liberase (0.32 mg/ml) per spleen. Cut the spleens into small pieces and incubate at RT for 10–15 min. Isolate a spleen from unlabelled mice as a negative control.
3. Make a single-cell suspension by using a cell strainer and pressing with a 5 ml syringe plunger (see Subheading 3.1). Add 10×

volume of PBS containing 2% FCS. Harvest the splenocytes into a tube and pellet the cells by centrifugation.

4. Resuspend the cell pellet in ACK lysis buffer and leave at RT for 2 min. Stop the lysis reaction by adding 10× volume of PBS containing 2% FCS and pellet the cells by centrifugation.
5. Wash the cells with PBS containing 2% FCS. Pass the cells through a 40 μm strainer. Take an aliquot for cell counting. Pellet the cells again by centrifugation.
6. Resuspend cell pellet in PBS-EDTA buffer (10^8 cells/ml).
7. Incubate with Fc receptor blocker on ice for 10 min.
8. Incubate with 1 μg/ml biotinylated antibodies against CD3, Thy1.2, CD19, and TER119 on ice for 30 min.
9. Wash Streptavidin-BioMag beads 3× during antibody incubation. Use 750 μl beads/10^9 cells in 8 ml PBS-EDTA buffer.
10. Wash cells once by adding 20× volume of buffer, and centrifuge for 5 min.
11. Resuspend cells in PBS-EDTA buffer (10^8 cells/ml). Add to Streptavidin-BioMag beads. Incubate at 4–8 °C for 30 min. Invert every 10 min to mix.
12. Remove magnetic beads on the magnetic stand. Transfer cells unbound to the magnetic beads to a new tube. Repeat the same step 3 times. Pellet cells by centrifugation (see Note 6).
13. Count cells. Use 1×10^6 cells to continue staining cell surface antigens followed by intracellular staining of BrdU.
14. Cell surface staining:
    (a) Add 1×10^6 BrdU-labeled cells in 50 μl of blocking buffer to flow cytometry tubes, and incubate on ice for 10 min.
    (b) Prepare fluorescent antibody cocktail specific for (1) anti-CD11c-PE (1:100) and anti-CD8-PE-Cy5 (1:150) and anti-CD4-APC (1:100) or (2) anti-CD11c-APC and anti-CD11b-PE or (3) anti-CD11c-APC (1:100) and anti-PDCA-PE (1:10) in blocking buffer. Add 50 μl antibody cocktail per tube and mix well.
    (c) Incubate cells with antibodies on ice for 20 min.
    (d) Wash the cells once by adding 2 ml of staining buffer per tube, centrifuge for 5 min at 200–300 × *g*, and discard supernatant.
15. Fix and permeabilize cells. Perform BrdU-FITC staining according to the BrdU Flow kit instructions.
16. Resuspend the cells in 0.5 ml staining buffer per tube and analyze cells on a flow cytometer. Gate on cells with specific markers and analyze the percentage of $BrdU^+$ cells using a software such as Flowjo (Fig. 2).

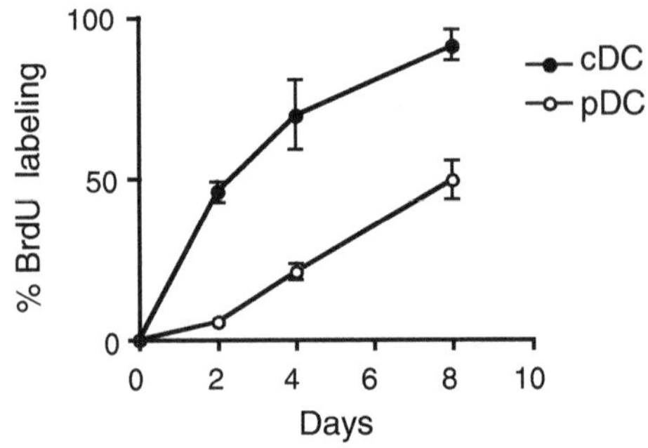

Fig. 2. Labeling of DCs with BrdU in vivo. Two-Days-old C57BL/6 mice were injected with BrdU on day 0 followed by feeding with BrdU in the drinking water for different days. Percentages of $BrdU^{+}$ cells among $CD11c^{+}CD11b^{+}$ conventional DC (cDC) or $CD11c^{low}PDCA\text{-}1^{+}$ plasmacytoid DC (pDC) were plotted.

4. Notes

1. Because DCs represent a small population of cell types in vivo, the purity of DCs isolated directly from lymphoid organs using magnetic beads depends on negative selection of other cell types. We have found that direct isolation of $CD11c^{+}$ DCs using anti-CD11c MACS-beads without negative selection yields DC with approximately 50% purity. After negative selection, >95% pure $CD11c^{+}$ DCs can be isolated using anti-CD11c MACS-beads.
2. Culturing monocytes from the bone marrow as described generally yields 70–80% $CD11c^{+}$ DCs. Further enrichment with anti-CD11c-MACS beads is important in certain experiments to achieve over 95% $CD11c^{+}$ DCs.
3. Some BMDCs attach to the wells and need to be flushed off the plates using a pipette for analyses. We have found that using 96-well U-bottom microtest polystyrene plates (BD Bioscience, Product Number: 351177) helps to reduce cell attachment and facilitate the harvesting of the cells. The plates can be sterilized under UV light for 10 min before use.
4. To assay T cell-mediated killing of DCs in vitro, T cells need to be activated in advance. In addition, the killing by $CD4^{+}$ or $CD8^{+}$ T effector cells is antigen dependent. It is important to pulse DCs with appropriate antigens before using them as target cells in kill assays.
5. Cytotoxicity mediated by other cell types, such as natural killer (NK) cells or polyclonal T regulatory (Treg) cells, will not require pulsing of DCs with antigens. However, these effector cells generally need to be activated to acquire killing activities toward DCs.

6. The half-life for DCs estimated by in vivo BrdU labeling is about 2 days for CD11c$^+$CD11b$^+$ myeloid DCs and 7–8 days for CD11c^{low}PDCA-1$^+$ plasmacytoid DCs (Fig. 2). However, DCs undergo much faster turnover in vitro (5). This suggests that local microenvironment plays a major role in regulating DC lifespan in vivo. Because different DC subsets represent small fractions of the total cell population, it is necessary to enrich DCs first by depleting other cell types before staining with anti-BrdU.
7. The procedures described here can be adapted to study dendritic cell death in different settings of immune responses, such as different phases of infections. The effects of various receptors expressed on DCs, and cytokines and chemokines produced in the microenvironment, may affect the lifespan of DCs at various stages of an immune response. The interplay between different cell types and soluble factors in regulating programmed cell death in DCs will be an important area for investigation.

Acknowledgements

This work was supported by NIH grants R01AI074949, R01AI056210 (J.W.), and R01DK083164 (M.C.).

References

1. Chen M, Wang YH, Wang Y, Huang L, Sandoval H, Liu YJ, Wang J (2006) Dendritic cell apoptosis in the maintenance of immune tolerance. Science 311:1160–1164
2. Stranges PB, Watson J, Cooper CJ, Choisy-Rossi CM, Stonebraker AC, Beighton RA, Hartig H, Sundberg JP, Servick S, Kaufmann G, Fink PJ, Chervonsky AV (2007) Elimination of antigen-presenting cells and autoreactive T cells by Fas contributes to prevention of autoimmunity. Immunity 26:629–41
3. Kamath AT, Henri S, Battye F, Tough DF, Shortman K (2002) Developmental kinetics and lifespan of dendritic cells in mouse lymphoid organs. Blood 100:1734–1741
4. O'Keeffe M, Hochrein H, Vremec D, Caminschi I, Miller JL, Anders EM, Wu L, Lahoud MH, Henri S, Scott B, Hertzog P, Tatarczuch L, Shortman K (2002) Mouse plasmacytoid cells: long-lived cells, heterogeneous in surface phenotype and function, that differentiate into CD8(+) dendritic cells only after microbial stimulus. J Exp Med 196:1307–1319
5. Chen M, Huang L, Shabier Z, Wang J (2007) Regulation of the lifespan in dendritic cell subsets. Mol Immunol 44:2558–2565

Chapter 7

Detection of Necrosis by Release of Lactate Dehydrogenase Activity

Francis Ka-Ming Chan, Kenta Moriwaki, and María José De Rosa

Abstract

Apoptosis and necrosis are two major forms of cell death observed in normal and disease pathologies. Although there are many assays for detection of apoptosis, relatively few assays are available for measuring necrosis. A key signature for necrotic cells is the permeabilization of the plasma membrane. This event can be quantified in tissue culture settings by measuring the release of the intracellular enzyme lactate dehydrogenase (LDH). When combined with other methods, measuring LDH release is a useful method for the detection of necrosis. In this chapter, we describe the step-by-step procedure for detection of LDH release from necrotic cells using a microtiter plate-based colorimetric absorbance assay.

Key words: Programmed necrosis, Necroptosis, Membrane leakage, Lactate dehydrogenase

1. Introduction

Necrosis is a type of cell death that is morphologically characterized by swelling and rupture of intracellular organelles, eventually leading to the breakdown of the plasma membrane (1–3). Plasma membrane leakage from necrotic cells causes the release of intracellular contents into the extracellular milieu. Therefore, necrotic cell death evokes inflammatory responses and is closely associated with inflammatory diseases (4). In contrast to apoptosis, necrosis was once considered as a passive and accidental form of cell death (5). However, several recent studies have clearly established that necrosis can be a programmed event, i.e., involves specific genetic pathways (6–8). Death cytokines in the tumor necrosis factor (TNF) family are potent triggers of necrosis. In this chapter, we describe a convenient, nonradioactive, and high-throughput method to detect necrosis, including secondary necrosis occurring after apoptosis, based on the leakage of a cytoplasmic enzyme, lactate dehydrogenase (LDH), from the damaged cells.

Andrew L. Snow and Michael J. Lenardo (eds.), *Immune Homeostasis: Methods and Protocols*, Methods in Molecular Biology, vol. 979, DOI 10.1007/978-1-62703-290-2_7, © Springer Science+Business Media New York 2013

Historically, necrotic cell death is evaluated by determining damage of the plasma membrane. The cytotoxicity assays for measuring necrosis are principally divided into two categories: those that are based on differential uptake of DNA-binding dyes that do not traverse the plasma membrane of living cells, such as propidium iodide, and those that are based on the leakage of intracellular molecules through compromised plasma membrane. LDH is a soluble cytoplasmic enzyme that is present in almost all cells and is released into the extracellular space when the plasma membrane is damaged (9). To detect the leakage of LDH into cell culture medium, a tetrazolium salt is used in this assay. In the first step, LDH produces reduced nicotinamide adenine dinucleotide (NADH) when it catalyzes the oxidation of lactate to pyruvate. In the second step, a tetrazolium salt (INT) is converted to a colored formazan product using newly synthesized NADH in the presence of an electron acceptor (10). The amount of formazan product can be colorimetrically quantified by standard spectroscopy. Because of the linearity of the assay, it can be used to enumerate the percentage of necrotic cells in a sample (see Note 1).

2. Materials

2.1. Equipment

1. Microtiter plate reader (e.g., FLUOstar Optima, BMG LABTECH).
2. Centrifuge with plate adaptors (e.g., Sorvall Legend RT).
3. Multichannel pipettes and repeat pipettor.

2.2. Reagents and Supplies

1. Flat-bottom or U-bottom 96-well microtiter plate.
2. RPMI 1640, Dulbecco's Modified Eagle Medium (DMEM), or any other media depending on the cell type used (see Notes 2–4).
3. Fetal bovine serum (FBS, multiple vendors) (see Note 5).
4. 200 mM Tris Buffer, pH 8.0. Dissolve 24.2 g Tris base in 1 L Milli-Q water. Adjust pH to 8.0 with hydrochloric acid. Sterilize by autoclave.
5. 2× LDH assay buffer: Dissolve 223 mg INT (2-p-iodophenyl-3-p-nitrophenyl-5-phenyl tetrazolium chloride), 57 mg PMS (N-methylphenazonium methyl sulfate) (see Note 6), 575 mg NAD (nicotinamide adenine dinucleotide), and 3.2 g lactic acid in 480 mL 200 mM Tris buffer solution, pH 8.0 (see Notes 7–8). Store the 2× LDH assay buffer in small aliquots at −20°C (see Notes 9–10).
6. 10× Lysis solution: 9% Triton X-100. Dissolve 9 mL Triton X-100 with 91 mL Milli-Q water.
7. Stop solution: 1 M acetic acid (see Note 11).

3. Methods

All the reactions should be performed at room temperature (22–25°C). Warm all the components to 22–25°C before starting the experiment.

3.1. Cell Culture Preparation (see Note 12)

1. Seed cells in a 96-well flat-bottom microtiter plate at a density of 1×10^4–5×10^4 cells/well in 100 μL of culture medium (e.g., RPMI or DMEM + <5% FBS). Samples in each experimental group should be prepared in triplicate wells (see Notes 13, 14).
2. On the same plate, prepare the following controls in triplicates:
   (a) Maximum LDH release—Seed the same number of cells in these wells as in step 1.
   (b) Medium alone—These wells contain only medium but no cells.
   (c) Volume correction control—These wells contain only medium but no cells.
3. For adherent cells, allow the cells to incubate for 12–16 h prior to treatment. For suspension cultures, treatment can be initiated when seeding of cells is completed.
4. Culture the cells for the required amount of time in a humidified 37°C incubator equilibrated with 5% CO_2.

3.2. Harvest Supernatants

1. To the maximum LDH release wells and the volume correction control wells (see Subheading 3.1, step 2), add 10 μL of 10× lysis solution.
2. Return the plate to a humidified 37°C, 5% CO_2 incubator for 45 min.
3. Centrifuge the microtiter plate in a centrifuge at $250 \times g$ for 5 min (see Note 15). It is advantageous to seal the plate with Parafilm if it can be accommodated by the carrier.
4. With a multichannel pipette, collect 50 μL of culture supernatant from each well and transfer it to a new microtiter plate (see Notes 16–18). Be careful not to transfer any cellular material from the bottom of the well.

3.3. LDH Measurement

1. Add 50 μL of the reconstituted 2× LDH assay buffer to the supernatants. Mix the reagents by gentle shaking for 30 s (see Notes 19, 20).
2. Protect the assay plate from light. Incubate the plate at room temperature for 10–30 min (see Note 21).
3. Add 50 μL of stop solution to the wells. Mix the reagent by shaking for 30 s (see Note 22).
4. Measure absorbance between 490 and 520 nm (see Note 23).

3.4. Data Analysis

1. The reading from the untreated and test wells is subtracted from the reading of the "medium alone" control wells. This is the corrected reading for the untreated or test well.
2. Subtract the reading of "volume correction control" wells from the reading of the "maximum LDH release" wells. This is the corrected maximum release reading.
3. Calculate the percentage of cytotoxicity using the following formula:
   % Cytotoxicity = 100 × (corrected reading from test well − corrected reading from untreated well)/(corrected maximum LDH release control − corrected reading from untreated well).

4. Notes

1. It is noteworthy that the LDH release assay described here does not distinguish between primary necrosis and secondary necrosis as a consequence of apoptotic cell death. Other assays, such as those that measure caspase activity or DNA strand breaks in apoptotic cells, can be performed to distinguish whether any detected LDH release is due to secondary necrosis from apoptotic cells.
2. Phenol red is a common pH indicator included in tissue culture media. Phenol red-free medium should be used to minimize interference with the absorbance measurement.
3. Pyruvate is a common culture media supplement. However, because it is an intermediate product of the LDH reaction, it should NOT be included in culture media to avoid interference of the LDH reaction.
4. INT can be reduced nonenzymatically by strong reducing agents. Hence, reducing agents such as ascorbate, 2-mercaptoethanol, and dithiothreitol should be excluded from the culture medium.
5. Animal serum contains different quantities of LDH. The amount of LDH present in the serum varies from batch to batch. In general, human AB serum is relatively low in LDH activity, whereas FBS contains relatively high levels of LDH. We therefore recommend that either human serum or no more than 5% FBS should be used to reduce background LDH activity.
6. PMS is an electron carrier for the second step of the LDH reaction. Instead of PMS, other electron carriers such as Mendola blue or diaphorase can also be used.
7. The chemicals may not completely dissolve immediately. Care should be taken to completely dissolve all chemicals prior to assay development by agitation.

8. Commercial vendors sometimes provide the components in a lyophilized or concentrated form. Follow the manufacturer's instructions for reconstitution of the components to ensure the optimal LDH assay performance.
9. Exposure to ambient light can degrade the LDH assay buffer substrates and give rise to low absorbance values. Protect the aliquots from light.
10. Repeated freezing and thawing of the LDH assay buffer should be avoided. We recommended storing aliquots of the buffer in quantities sufficient for a single assay experiment at −20°C (e.g., 5 mL for one 96-well microtiter plate).
11. Instead of 1 M acetic acid, 1–5% SDS or other reagents such as strong base (1 N NaOH) or acid (1 N HCl) can be used to stop the reaction.
12. The LDH protocol described here can be adapted to measurement of cell-mediated cytotoxicity (11).
13. Different cell types contain different amounts of LDH. Therefore, the optimal cell density used should be determined empirically by performing a "total lysis" using different numbers of cells. The resulting LDH absorbance reading can be plotted against the cell number. The optimal cell number should be the one that falls within the linear range of the absorbance curve.
14. Because LDH can "build up" in the culture media after overnight incubation and contribute to high-background LDH activity, we recommend replacing the old media with fresh media prior to induction of cell death for adherent cell cultures.
15. We recommend centrifugation of the cells to minimize inadvertent inclusion of cellular material during the harvest of the culture supernatants. However, care should be taken to minimize physical damage of the cells during centrifugation, which can lead to increased background "leakage" of LDH into the culture medium.
16. A smaller volume of culture medium can be harvested depending on the sensitivity of the assay. Factors such as the number of cells used, the amount of LDH expressed in a given cell type, and the reagents used can affect assay sensitivity. For 100 μL culture on 96-well flat-bottom microtiter plates, we recommend that no more than 50 μL be harvested to minimize contamination of the culture supernatant with cellular material.
17. After harvest, cell-free culture supernatants can be stored at 2–8°C for a few days without appreciable loss of LDH activity.
18. After harvest of culture supernatants, the cells in the original plate can be used for other complementary cell viability assays such as MTS (3-(4,5-dimethylthiazol-2-yl)-5-(3-carboxymethoxyphenyl) -2-(4-sulfophenyl)-2H-tetrazolium) (12).

19. To minimize the variation in incubation time between different wells, especially when a large number of sample wells is involved, use a multichannel pipette to dispense the 2× LDH assay buffer to the assay wells.
20. An optional positive control of LDH standard can be included.
21. The reaction time can be decreased or increased depending on the color development. Monitor the color conversion periodically. Stop the reaction before the color conversion of the sample wells reaches the level of that seen in the maximum LDH release wells.
22. Measure the absorbance within one hour after addition of the stop solution. If the rate of the reaction is slow (i.e., slow color conversion), absorbance can be measured again without adding the stop solution.
23. Air bubbles present in wells affect the absorbance readings. Air bubbles can be purged using a needle.

Acknowledgements

This work was supported by NIH grants (AI083497 and AI088502) to F. K-M. C. K. M. is supported by a postdoctoral fellowship from the Uehara Memorial Foundation, Japan. MJDR is supported by a CONICET fellowship from Argentina.

References

1. Schweichel JU, Merker HJ (1973) The morphology of various types of cell death in prenatal tissues. Teratology 7:253–266
2. Moquin D, Chan FK (2010) The molecular regulation of programmed necrotic cell injury. Trends Biochem Sci 35:434–441
3. Challa S, Chan FK (2010) Going up in flames: necrotic cell injury and inflammatory diseases. Cell Mol Life Sci 67:3241–3253
4. Kono H, Rock KL (2008) How dying cells alert the immune system to danger. Nat Rev Immunol 8:279–289
5. Chipuk JE, Green DR (2005) Do inducers of apoptosis trigger caspase-independent cell death? Nat Rev Mol Cell Biol 6:268–275
6. Chan FK, Shisler J, Bixby JG, Felices M, Zheng L, Appel M, Orenstein J, Moss B, Lenardo MJ (2003) A role for tumor necrosis factor receptor-2 and receptor-interacting protein in programmed necrosis and antiviral responses. J Biol Chem 278:51613–51621
7. Cho YS, Challa S, Moquin D, Genga R, Ray TD, Guildford M, Chan FK (2009) Phosphorylation-driven assembly of the RIP1-RIP3 complex regulates programmed necrosis and virus-induced inflammation. Cell 137:1112–1123
8. Holler N, Zaru R, Micheau O, Thome M, Attinger A, Valitutti S, Bodmer JL, Schneider P, Seed B, Tschopp J (2000) Fas triggers an alternative, caspase-8-independent cell death pathway using the kinase RIP as effector molecule. Nat Immunol 1:489–495
9. Burd JF, Usategui-Gomez M (1973) A colorimetric assay for serum lactate dehydrogenase. Clin Chim Acta 46:223–227
10. Korzeniewski C, Callewaert DM (1983) An enzyme-release assay for natural cytotoxicity. J Immunol Methods 64:313–320
11. Decker T, Lohmann-Matthes ML (1988) A quick and simple method for the quantitation of lactate dehydrogenase release in measurements of cellular cytotoxicity and tumor necrosis factor (TNF) activity. J Immunol Methods 115:61–69
12. Cory AH, Owen TC, Barltrop JA, Cory JG (1991) Use of an aqueous soluble tetrazolium/formazan assay for cell growth assays in culture. Cancer Commun 3:207–212

Chapter 8

Assessment of CD4⁺ and CD8⁺ T Cell Responses Using MHC Class I and II Tetramers

Sema Kurtulus and David Hildeman

Abstract

The low frequency of T cells specific for given antigens makes the study of antigen-specific T cell responses difficult. The development of MHC class I and II tetramer staining techniques allows precise quantification and tracking of antigen-specific CD8⁺ and CD4⁺ T cell responses. Here, we describe a protocol for MHC class I and II tetramer staining of mouse T cells isolated from various tissues of mice infected with lymphocytic choriomeningitis virus (LCMV) or with murine cytomegalovirus (MCMV).

Key words: MHC Class I tetramer staining, MHC-peptide complexes, CD8⁺ T cells, CD4⁺ T cells, D^b gp33, D^b M45, I-A^b gp61, I-A^b M25, Flow cytometry, LCMV, MCMV

1. Introduction

For many years, analysis of antigen-specific T cell responses was hampered by their lack of identifying markers. Many of the methods used by immunologists to measure antigen-specific responses have important limitations. Limiting dilution assays and cytotoxicity assays were some of the first tools used to measure bulk T cell responses (1–4). While limiting dilution assays are fairly quantitative, they depend largely on expansion, survival, and subsequent function of precursors under particular culture conditions. On the other hand, cytotoxicity assays measure function at a population level, and quantitation is difficult and often lacks precision. TCR transgenic (Tg) mice and adoptive transfer approaches were subsequently developed and can be helpful (5), but nonphysiologically high levels of TCR Tg cells can lead to phenotypic abnormalities in these cells (6). Another technique, intracellular cytokine staining, can identify antigen-specific responding T cells; however, this approach is limited by the cytokine-producing ability of the T cell. The development of MHC "tetramers", recombinant, multimeric,

Andrew L. Snow and Michael J. Lenardo (eds.), *Immune Homeostasis: Methods and Protocols*, Methods in Molecular Biology, vol. 979, DOI 10.1007/978-1-62703-290-2_8, © Springer Science+Business Media New York 2013

MHC molecules with attached peptides, has revolutionized our ability to track and quantify antigen-specific T cell responses ex vivo (7). The quadrivalent nature of the MHC/peptide tetramers allows for sufficient avidity to interact and stably bind to T cell receptors to visualize specific T cells by flow cytometry. Indeed, this approach uncovered the vastly underestimated response reported by previous nonquantitative methods (8).

To generate tetramers, MHC class I α-chains and β2-microglobulin are first produced as recombinant proteins in bacteria and then refolded in vitro in the presence of the peptide of interest into MHC/peptide monomers (7, 9). Engineering of a BirA recognition site into the class I α-chain allows for site-specific, single biotinylation by incubation of MHC/peptide monomers. Biotinylated monomers can be flash frozen indefinitely in aliquots that can be thawed and coupled to fluorochrome-labeled streptavidin to generate MHC tetrameric staining reagents.

Unlike MHC class I–peptide complexes that fold together in vitro, mouse MHC class II–peptide complexes are generally assembled in insect cells as preassembled monomers (10, 11). This is because mouse MHC class II molecules are difficult to fold in vitro and the covalent, genetic linkage of peptides to MHC class II β-chains allows for proper folding and assembly in insect cells (10, 11). Similar to class I MHC molecules, purified monomeric MHC class II–peptide complexes are singly biotinylated on a BirA recognition site in the MHC class II β-chain. Biotinylated monomers are assembled into tetramers by coupling with fluorochrome-linked streptavidin.

MHC class I and II tetramer staining can be used to quantify antigen-specific T cells in various mouse tissues such as spleen, lymph nodes (12), and peripheral tissues as livers, gut, kidney (13), and brain (14). After lymphocytic choriomeningitis virus (LCMV) infection of C57BL/6 mice, a major immunodominant epitope recognized by $CD8^+$ T cells is derived from the glycoprotein (GP) peptide amino acids 33–41 presented by MHC class I $H\text{-}2D^b$ molecules, while a major $CD4^+$ T cell response is directed against another GP epitope (amino acids 61–80) presented by $I\text{-}A^b$ (15, 16). After murine cytomegalovirus (MCMV) infection of C57BL/6 mice, a major immunodominant epitope recognized by $CD8^+$ T cells is the MCMV E protein M45 epitope (HGIRNASFI) presented by $H\text{-}2D^b$; while another M protein epitope M25 (amino acids 409–423) is presented by $I\text{-}A^b$ and recognized by $CD4^+$ T cells (17, 18). In this chapter, we will describe a simple protocol for isolation of cells from tissues and their staining and analysis using MHC class I and II tetramers. Further, we will demonstrate the non-cross-reactivity of $CD8^+$ and $CD4^+$ T cells by staining cells from LCMV- versus MCMV-infected mice using LCMV- and MCMV-specific MHC tetramers.

2. Materials

2.1. Infection of Mice

1. Armstrong-3 strain of LCMV: LCMV are grown in BHK-21 cells, and the number of plaque-forming units (pfu) is assayed on Vero cells (9).
2. Smith strain of MCMV: MCMV are prepared by in vivo propagation in Balb/C mice; salivary gland homogenates are prepared 14 days postinfection, and viral titer in plaque-forming units (pfu) is measured by plaque assays on mouse embryo fibroblasts (19, 20).

2.2. Preparation of Cell Suspensions from Tissues

1. 20× Balanced salt solution (BSS): 5.6 mM glucose, 0.4 mM KH_2PO_4, 1.3 mM Na_2HPO_4, 1.3 mM $CaCl_2 \cdot 2H_2O$, 5.4 mM KCl, 137 mM NaCl, 0.8 mM $MgSO_4$, 1 mM $MgCl_2 \cdot 6H_2O$, and 0.001% phenol red. Mix and filter sterilize (see Note 1).
2. ACK lysis buffer: Prepare 155 mM NH_4Cl buffer, adjust the pH to 7.5, and then add $KHCO_3$ at a final concentration of 10 mM and phenol red 0.0005%. Filter sterilize.
3. CTM: Minimum essential medium (S-MEM) complemented with 2 mM L-glutamine, 0.1 mM nonessential amino acids, 0.5× essential amino acid mixture, 1 mM sodium pyruvate, 100 U/ml penicillin, 100 μg/ml streptomycin sulfate, 50 μg/ml gentamicin, 50 μM β-mercaptoethanol, 3 mM dextrose, and 10% Fetal Bovine Serum (FBS) (see Note 2).
4. Click's medium (EHAA) (available from multiple vendors).
5. 5.1 mg/ml Liberase +0.8 mg/ml DNAse I solution in EHAA medium.
6. 0.1 M EDTA.
7. 37.5% isotonic Percoll.
8. 70 μm mesh cell strainers, 50 ml tubes, 96-well plates.
9. Tabletop centrifuge.

2.3. Staining for Flow Cytometry

1. Flow buffer: BSS supplemented with 2% FBS and 0.1% w/v sodium azide (see Note 3).
2. Paraformaldehyde (PFA) solution (2% w/v) in Phosphate Buffered Saline (PBS) (see Note 4).
3. MHC class I tetramers: D^b–gp33 or D^b–M45 coupled with SA-PE (see Note 5).
4. MHC class II tetramer: I-A^b–gp61 or I-A^b–M25 coupled with SA-PE.
5. Antibodies against mouse CD4 and CD8 and/or other surface markers, coupled to desired fluorochromes other than PE (multiple vendors).
6. Flow tubes, flow cytometry analyzer.

3. Methods

3.1. Infection with LCMV

Inject mice with 2×10^5 pfu of LCMV or with 1×10^5 pfu of MCMV intraperitoneally. Allow for expansion of T cells for at least 5–6 days (see Note 6). Specific T cells can be stained with MHC tetramers for the lifetime of the infected animal.

3.2. Preparation of Single Cell Suspensions from Mouse Spleen

1. Harvest the spleen of the infected mouse, and crush through a 70 μm mesh cell strainer into 50 ml tube to generate a single cell suspension.
2. Add 35 ml of BSS through the strainer to allow passage of the cells into the tube.
3. Spin the cells down at $300 \times g$ for 5 min at 4 °C. Discard the supernatant.
4. To lyse red blood cells, add 1.5 ml of ACK lysis buffer, and incubate for 2.5 min at room temperature (RT).
5. Increase volume to 50 ml by adding BSS and centrifuge at $300 \times g$ for 5 min, at 4 °C.
6. Discard the supernatant.
7. Repeat steps 5 and 6 to wash one more time (see Note 7).
8. Resuspend the cells in 3 ml of CTM.
9. Count the cells and adjust concentration to add 2×10^6 cells per well of a 96-well plate.

3.3. Preparation of Single Cell Suspensions from Mouse Liver

1. Harvest liver tissue from the mouse into 5 ml of serum-free EHAA medium and mince into small pieces.
2. Add 50 μl of Liberase + DNAse I solution on the liver pieces in a 50 ml conical tube, and incubate at 37 °C for 30 min, swirling every 5 min.
3. Add 2 ml of 0.1 M EDTA to a final concentration of 0.03 M.
4. Filter liver tissue through cell strainers, and add BSS up to 35 ml (see Note 8).
5. Centrifuge cells at $300 \times g$ for 5 min at 4 °C. Discard the supernatant.
6. Resuspend cells with 25 ml of 37.5% isotonic Percoll.
7. Centrifuge cells at $690 \times g$ for 12 min. Pipet the supernatants off the cell pellet carefully.
8. Add 50 ml BSS and wash the cells by repeating step 5.
9. Lyse the red cells by following the steps 4–9 of Subheading 3.2.

3.4. MHC Class I Tetramer Staining

1. Add 100 μl of flow buffer onto 2×10^6 cells seeded in wells.
2. Centrifuge the cells down at $300 \times g$ for 5 min at 4 °C. Discard the supernatant.
3. Add appropriate amount of D^b–gp33 or D^b–M45 coupled with SA-PE in 70 μl of flow buffer/well (see Note 9).
4. Incubate the cells with the tetramer for 45 min at 4 °C.
5. Prepare antibody mixtures against CD8, CD16/32, and other surface molecules of interest in 20 μl of BSS/well. Add on top of each well, without washing away the tetramer stain, and incubate for an additional 45 min at 4 °C (see Note 10).
6. Add 150 μl of flow buffer/well and centrifuge the cells down at $300 \times g$ for 5 min at 4 °C. Discard the supernatant.
7. Add 200 μl of flow buffer/well and centrifuge the cells down at $300 \times g$ for 5 min at 4 °C. Discard the supernatant.
8. Fix the cells by adding 100 μl of 2% PFA for 20 min at RT.
9. Add 150 μl of flow buffer/well and centrifuge the cells down at $300 \times g$ for 5 min at 4 °C. Discard the supernatant.
10. Resuspend the cells with 200 μl of flow buffer/well, and analyze on a flow cytometer.

3.5. MHC Class II Tetramer Staining

1. Add 100 μl of CTM onto 2×10^6 cells seeded in wells.
2. Centrifuge the cells at $300 \times g$ for 5 min at 4 °C. Discard the supernatant.
3. Add appropriate amount of I-A^b–gp61 coupled with SA-PE in 70 μl of CTM/well.
4. Incubate the cells with the tetramer for 75 min at 37 °C.
5. Prepare antibody mixtures against CD4, CD16/32, and other surface molecules of interest in 20 μl of CTM/well. Add on top of each well, without washing away the tetramer stain, and incubate for an additional 45 min at 37 °C (see Note 11).
6. Add 150 μl of CTM/well and centrifuge the cells at $300 \times g$ for 5 min at 4 °C. Discard the supernatant.
7. Fix the cells by adding 100 μl of 2% PFA for 20 min at RT.
8. Add 150 μl of flow buffer/well and centrifuge the cells at $300 \times g$ for 5 min at 4 °C. Discard the supernatant.
9. Resuspend the cells with 200 μl of flow buffer/well, and analyze on a flow cytometer.

4. Notes

1. When preparing BSS, two separate solutions should be prepared to prevent precipitation. Glucose, KH_2PO_4, and Na_2HPO_4 should be dissolved together. $CaCl_2{\cdot}2H_2O$, KCl, 137 mM NaCl, $MgSO_4$, and $MgCl_2{\cdot}6H_2O$ should be prepared in a separate solution and should be mixed afterwards.
2. Supplements for the CTM media except for FBS can be prepared as a 100× stock solution and stored at −20 °C.
3. Sodium azide is very toxic if ingested or inhaled. Avoid contact with skin, eyes, or clothing.
4. Prepare a stock 4% w/v PFA solution. First heat deionized distilled H_2O (ddH_2O) until bubbles form, and then decrease the heat to 60 °C. Dissolve PFA in the water which takes around 10 min. If PFA does not dissolve, add 1–2 drops of NaOH into the mixture. After PFA dissolves, cool the solution down, and add 10× PBS to a final concentration of 1×. Adjust the pH to 7.6. Complete to the appropriate volume and store at −20 °C. If desired, a 10× stock PBS solution can be prepared (137 mM NaCl, 2.7 mM KCl, 100 mM Na_2HPO_4, and 2 mM KH_2PO_4), with pH adjusted to 7.2. Dilute 4% PFA to 2% PFA with ddH_2O.
5. D^b–gp33 or D^b–M45 monomers can be flash frozen in a dry ice/ethanol bath and stored at −80 °C, but once they are conjugated to fluorochromes, they are kept at +4 °C in the dark like other fluorochrome-conjugated antibodies.
6. Typically, endogenous T cell responses to acute viral infections can be quantified as early as 5 days after infection. However, peak $CD8^+$ T cell numbers occur between days 8 and 10 after infection. In the next 10–15 days after this point, most of the T cells undergo apoptotic cell death (12); however, sufficient numbers of detectable cells persist for the lifetime of the animal. Depending upon the markers to be interrogated on antigen-specific T cells, it might be useful to collect 1–2 million events on the flow cytometer. This may mean scaling up the staining of 4–6 million cells per sample.
7. Do not skip the second wash as this could affect cell viability.
8. We find it useful to crush the liver pieces over the cell strainer to get the cells through the strainer.
9. The appropriate amount of the tetramer is determined by comparing several dilutions of the tetramer staining in infected cells to uninfected cells or to a stain with an irrelevant tetramer. The dilutions that have substantial staining on infected cells but having the lowest background in uninfected cells are chosen

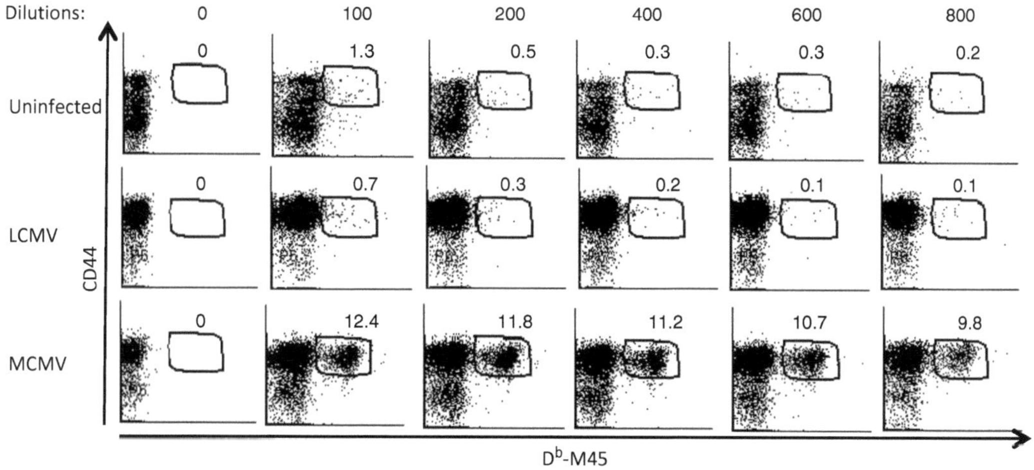

Fig. 1. Determination of appropriate MHC tetramer concentration to use for staining. Representative dot plots show staining with various dilutions of D^b–M45 tetramer and antibody against CD44 after gating on CD8⁺ T cells. Splenocytes from uninfected naïve C57/BL6 mice and mice infected with LCMV or with MCMV are stained with D^b–M45 tetramer at the dilutions indicated. Specific D^b–M45⁺ CD8⁺ CD44⁺ T cells are only detected in MCMV-infected mice, and their percentages within total CD8⁺ T cells are indicated on plots. The appropriate dilution to use the tetramer is 1:400 dilution because the percentage of D^b–M45⁺ CD44⁺ cells within CD8⁺ T cells is high, the mean fluorescence intensity (MFI) of the D^b–M45 stain within the tetramer+population remains high, and the nonspecific staining in naïve and LCMV-infected mice are low at this concentration.

(Fig. 1). As an independent test to determine whether or not MHC tetramers stain activated cells, we infected mice with either MCMV or LCMV and assessed staining of splenocytes from each mouse with both MCMV and LCMV tetramers. Results show fine specificity of both class I and class II MHC tetramers (Fig. 2).

10. CD16/32 refers to the murine Fcγ receptors. In flow staining, these receptors are generally blocked with antibodies to avoid nonspecific staining. Anti-CD16/32 antibody is produced by the 2.4G2 hybridoma (available from ATCC) and can be purified from 2.4G2 cell culture supernatant using a protein G column.

11. For class II staining, we include anti-CD16/32 antibody that is conjugated to a fluorochrome. $CD16/32^+$ cells will be excluded from $CD4^+$ gate to eliminate $CD4^+$ myeloid cells from the analysis. We have found that CD16/32 is not expressed on $CD4^+$ T cells and that cells bearing Fcγ receptors nonspecifically stain with MHC class II tetramers produced in baculoviral systems. Therefore, when analyzing the stains, a strategy that gates out $CD16/32^+$ events helps to decrease the background (Fig. 3).

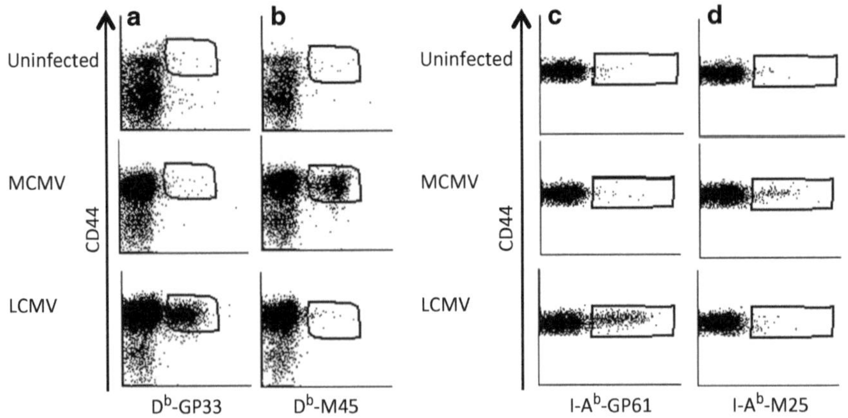

Fig. 2. Specificity of MHC class I and class II tetramers. Single cell suspensions from spleens of C57BL/6 mice uninfected or infected with either LCMV or MCMV were stained with antibodies against CD8 and CD44 and with D^b–GP33 (**a**) or D^b–M45 (**b**) tetramers. (**a**, **b**) Representative dot plots show splenocytes stained with antibody against CD44 and with D^b–GP33 (**a**) or D^b–M45 (**b**) tetramer after gating on $CD8^+$ T cells in naïve, LCMV- or MCMV-infected mice. (**c**, **d**) Splenocytes from each mouse are stained with antibodies against CD4 and CD16/32 and with I-A^b–GP61 (**c**) or I-A^b–M25 class II tetramers. Dot plots show CD4 and tetramer staining after gating on $CD4^+$ $CD16/32^-$ cells.

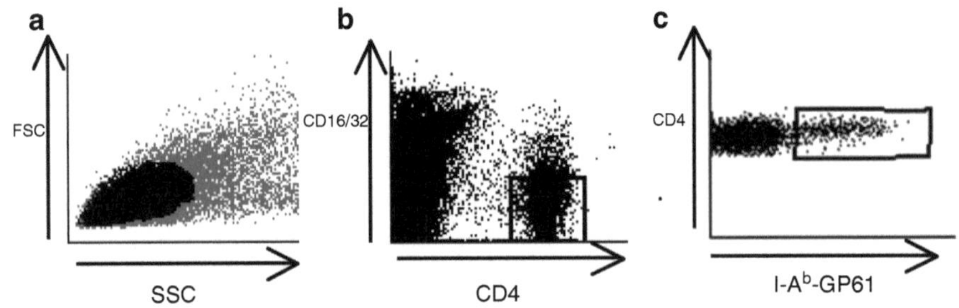

Fig. 3. Gating strategy for MHC class II tetramer staining. Representative dot plots showing the flow cytometry analysis of staining of splenocytes from LCMV-infected mice with I-A^b–GP61 tetramer and antibodies against CD4 and CD16/32. (**a**) Dot plot showing Forward Scatter (FSC) and Side Scatter (SSC) of stained cells. Plot area shown in black indicates the live gate. (**b**) Dot plot shows CD4 by CD16/32 staining on gated live cells. (**c**) Dot plot shows CD4 staining by I-A^b–GP61 tetramer staining on $CD4^+$CD16/32-gated events in (**b**).

Acknowledgments

This work was supported by NIH-Public Health Service Grants AI057753, DK081175 (D.A.H.). The authors also thank Drs. Kasper Hoebe and Rhonda Cardin for their assistance with MCMV infection.

References

1. Lau LL, Jamieson BD, Somasundaram T, Ahmed R (1994) Cytotoxic T-cell memory without antigen. Nature 369:648–652
2. Razvi ES, Welsh RM, McFarland HI (1995) In vivo state of antiviral CTL precursors. Characterization of a cycling cell population containing CTL precursors in immune mice. J Immunol 154:620–632
3. Selin LK, Nahill SR, Welsh RM (1994) Cross-reactivities in memory cytotoxic T lymphocyte recognition of heterologous viruses. J Exp Med 179:1933–1943
4. Tripp RA, Hou S, McMickle A, Houston J, Doherty PC (1995) Recruitment and proliferation of CD8+ T cells in respiratory virus infections. J Immunol 154:6013–6021
5. Butz EA, Bevan MJ (1998) Massive expansion of antigen-specific CD8+ T cells during an acute virus infection. Immunity 8:167–175
6. Marzo AL, Klonowski KD, Le Bon A, Borrow P, Tough DF, Lefrancois L (2005) Initial T cell frequency dictates memory CD8+ T cell lineage commitment. Nat Immunol 6:793–799
7. Altman JD, Moss PA, Goulder PJ, Barouch DH, McHeyzer-Williams MG, Bell JI, McMichael AJ, Davis MM (1996) Phenotypic analysis of antigen-specific T lymphocytes. Science 274:94–96
8. Murali-Krishna K, Altman JD, Suresh M, Sourdive DJ, Zajac AJ, Miller JD, Slansky J, Ahmed R (1998) Counting antigen-specific CD8 T cells: a reevaluation of bystander activation during viral infection. Immunity 8: 177–187
9. Fuller MJ, Zajac AJ (2003) Ablation of CD8 and CD4 T cell responses by high viral loads. J Immunol 170:477–486
10. Crawford F, Kozono H, White J, Marrack P, Kappler J (1998) Detection of antigen-specific T cells with multivalent soluble class II MHC covalent peptide complexes. Immunity 8: 675–682
11. Rees W, Bender J, Teague TK, Kedl RM, Crawford F, Marrack P, Kappler J (1999) An inverse relationship between T cell receptor affinity and antigen dose during CD4(+) T cell responses in vivo and in vitro. Proc Natl Acad Sci U S A 96:9781–9786
12. Wojciechowski S, Jordan MB, Zhu Y, White J, Zajac AJ, Hildeman DA (2006) Bim mediates apoptosis of CD127 (lo) effector T cells and limits T cell memory. Eur J Immunol 36: 1694–1706
13. Masopust D, Vezys V, Marzo AL, Lefrancois L (2001) Preferential localization of effector memory cells in nonlymphoid tissue. Science 291:2413–2417
14. Lin AA, Tripathi PK, Sholl A, Jordan MB, Hildeman DA (2009) Gamma interferon signaling in macrophage lineage cells regulates central nervous system inflammation and chemokine production. J Virol 83:8604–8615
15. Gairin JE, Mazarguil H, Hudrisier D, Oldstone MB (1995) Optimal lymphocytic choriomeningitis virus sequences restricted by H-2Db major histocompatibility complex class I molecules and presented to cytotoxic T lymphocytes. J Virol 69:2297–2305
16. Oxenius A, Bachmann MF, Ashton-Rickardt PG, Tonegawa S, Zinkernagel RM, Hengartner H (1995) Presentation of endogenous viral proteins in association with major histocompatibility complex class II: on the role of intracellular compartmentalization, invariant chain and the TAP transporter system. Eur J Immunol 25:3402–3411
17. Gold MC, Munks MW, Wagner M, Koszinowski UH, Hill AB, Fling SP (2002) The murine cytomegalovirus immunomodulatory gene m152 prevents recognition of infected cells by M45-specific CTL but does not alter the immunodominance of the M45-specific CD8 T cell response in vivo. J Immunol 169:359–365
18. Arens R, Wang P, Sidney J, Loewendorf A, Sette A, Schoenberger SP, Peters B, Benedict CA (2008) Cutting edge: murine cytomegalovirus induces a polyfunctional CD4 T cell response. J Immunol 180:6472–6476
19. Tabeta K, Georgel P, Janssen E, Du X, Hoebe K, Crozat K, Mudd S, Shamel L, Sovath S, Goode J, Alexopoulou L, Flavell RA, Beutler B (2004) Toll-like receptors 9 and 3 as essential components of innate immune defense against mouse cytomegalovirus infection. Proc Natl Acad Sci U S A 101:3516–3521
20. Orange JS, Wang B, Terhorst C, Biron CA (1995) Requirement for natural killer cell-produced interferon gamma in defense against murine cytomegalovirus infection and enhancement of this defense pathway by interleukin 12 administration. J Exp Med 182:1045–1056

Chapter 9

Homeostatic Proliferation of Mature T Cells

Christopher E. Martin, Kwesi Frimpong-Boateng, Darina S. Spasova, John C. Stone, and Charles D. Surh

Abstract

Under normal circumstances, the secondary lymphoid tissues contain a predictable number of T cells with a diverse T cell receptor (TCR) repertoire. Such a T cell pool must be of sufficient size to confer maximum protection of the host from infectious pathogens and cancer, but small enough not to overburden the host. The T cell pool is maintained by a combination of de novo T cell production by the thymus and by the long-term survival and gradual turnover of mature T cells in the periphery. The latter process, termed homeostatic proliferation, has been intensely investigated over the past 20 years, and a few techniques have been developed to facilitate these studies. In this chapter, we describe the experimental procedures that allow conspicuous visualization of homeostatic proliferation, which have been instrumental in facilitating recent advances in the study of T cell homeostasis.

Key words: CFSE, CellTrace Violet, Cell Proliferation Dye eFluor 670, T cell proliferation, CD45, CD90, Lymphopenic mice

1. Introduction

As for many cell types, the longevity of mature T cells is dependent on the continuous reception of survival signals from other cells in their environment. Accordingly, the size of the T cell pool is regulated by a limited supply of these survival factors, which have been defined by a few key experimental methods. We outline one widely used procedure, which enables the researcher to measure the T cell response to increased availability of survival factors under in vivo conditions. This method involves adoptive transfer of dye-labeled T cells into syngeneic lymphopenic hosts to observe the donor T cell proliferative response, which occurs in the absence of intentional immunization. Since the transferred T cells proliferate in an attempt to restore the depleted T cell pool to homeostasis, the phenomenon has been termed *homeostatic proliferation* and represents an accelerated model

Andrew L. Snow and Michael J. Lenardo (eds.), *Immune Homeostasis: Methods and Protocols*, Methods in Molecular Biology, vol. 979, DOI 10.1007/978-1-62703-290-2_9,

of normal T cell turnover (1). With this basic template, researchers have identified essential survival factors by testing whether addition or removal of individual candidate factors results in perturbation of homeostatic proliferation. Moreover, such studies on various populations of T cells have led to the elucidation of factors that support survival of nearly all subsets of mature T cells, including, naïve and memory αβ T cells, γδ T cells, NKT, and even NK cells (1–13). Our discussion is mostly limited to αβ T cells.

1.1. Homeostatic Factors

The important survival signals described to date for mature αβ T cell homeostasis include TCR signals from contact with self-peptides in major histocompatibility (MHC) molecules (self-pMHC ligands) and two cytokines belonging to the common-γ chain (CD132) family: IL-7 and IL-15. T cells are thought to have a mild affinity for self-pMHC ligands, such that this weak interaction is sufficient to drive low-level TCR signaling that is essential for survival. Naïve $CD4^+$ and $CD8^+$ T cells are especially reliant on contact with self-pMHC ligands for survival while memory T cells are able to do without it (1, 2). Homeostasis of all αβ T cells is critically dependent on IL-7 and IL-15. IL-7 is of central importance as its absence impairs survival of both naïve and memory T cells (3, 4). The related cytokine IL-15 plays an important role in inducing basal homeostatic proliferation of memory T cells and, to a lesser extent for survival of naïve $CD8^+$, but not $CD4^+$, T cells (4, 5, 14). Our current understanding of the survival factors that define the T cell niche has been covered comprehensively in multiple recent reviews (15–21).

1.2. Studying Homeostatic Proliferation

The study of homeostatic proliferation relies on many tools and technologies. Genetically modified mice and cellular dyes have proven to be two of the most essential resources for elegant and definitive experiments. While their use is not restricted to this field, there are special concerns that must be considered when selecting mice or the dye to observe homeostatic proliferation. We first discuss cellular dyes and novel reagents along with three types of lymphopenic hosts. More in-depth discussion of the variables that influence the choice of host and donor mice for this protocol is covered in Subheadings 3 and 4.

1.3. Dyes to Track Cell Division

The value of 5-(and 6-) carboxyfluorescein diacetate succinimidyl ester (abbreviated accurately as CFDA-SE, or popularly as CFSE) over other cellular dyes in proliferation assays is its ability to delineate progressive cell divisions with a high resolution (22–24). CFSE is considerably superior to membrane dyes, such as PKH26 (Sigma), or thio-reactive, chloromethyl derivative dyes (such as CellTracker™ Orange CMTMR or Green BIODIPY®, both from Invitrogen), which segregate less precisely between daughter cells, resulting in an imprecise definition of cell divisions (24, 25). When analyzed by flow cytometry, CFSE yields a series of clearly separated peaks, each representing a round of cell division (Fig. 1 top) (26, 27). Dyes that segregate imperfectly between daughter cells or have broader coefficient of variations yield histograms that more closely resemble a single mound or bell (Fig. 1 bottom). Beyond mere aesthetics, separated peaks enable reliable and accurate enumeration of the number of cell divisions and the population size of cells in each generation. While not recommended, similar data can be generated from a bell-shaped histogram of suboptimal cellular dyes through mathematical models that aim to approximate peaks beneath the mound. Software for such modeling is available within the FlowJo program (Tree Star, Inc) or ModFit LT (Verity Software House).

Two new amine-reactive cell proliferation probes have recently become available. CellTrace™ Violet (hereafter referred to as CTV, sold by Invitrogen) is excited by the 405 nm laser, and its emission is similar to that of Pacific Blue™. Cell Proliferation Dye eFluor® 670 (CPD670, available from eBioscience), is excited by the 633 nm laser and it is detected in the same channel on a flow cytometer as allophycocyanin. CTV and CPD670 were presumably developed for use with the myriad of genetically engineered GFP-reporter cell and mouse lines because CFSE is incompatible with GFP (i.e., both fluorophores have overlapping excitation and emission spectra). CTV is far superior to CPD670 in our hands and yields a similarly high resolution of cell divisions as seen with CFSE (Fig. 1). However, violet lasers are not commonly available in simple flow cytometer platforms and researchers at some institutes may be unable to utilize CTV. In such instances, those seeking to analyze proliferation of GFP-expressing cells may consider using PKH26 or CPD670. However, when labeling cells with PKH26 according to the manufacturer's instructions, cell death is an issue, as we often lose 50% of the cells after labeling. Conversely, this is not a concern with the labeling described below for CFSE, CTV, or CPD670. Thus, for experiments involving GFP^+ cells where a flow cytometer with a violet laser is not available, we recommend using CPD670 rather than PKH26. It should be noted that the labeling protocol described below was optimized for CFSE, not CPD670, making it slightly different from that published by the manufacturer for CPD670 (available at eBioscience.com). Thus, it remains possible, albeit unlikely, that a more optimal cell-labeling protocol for CPD670 could yield better resolution.

1.4. Lymphopenic Hosts

With respect to experiments in homeostatic proliferation, the commonly used lymphopenic hosts may be classified into three groups: acutely lymphopenic, chronically lymphopenic, and cytokine receptor deficient mice (Table 1). There are benefits and drawbacks associated with each type of lymphopenic host that should be considered when designing or evaluating experiments.

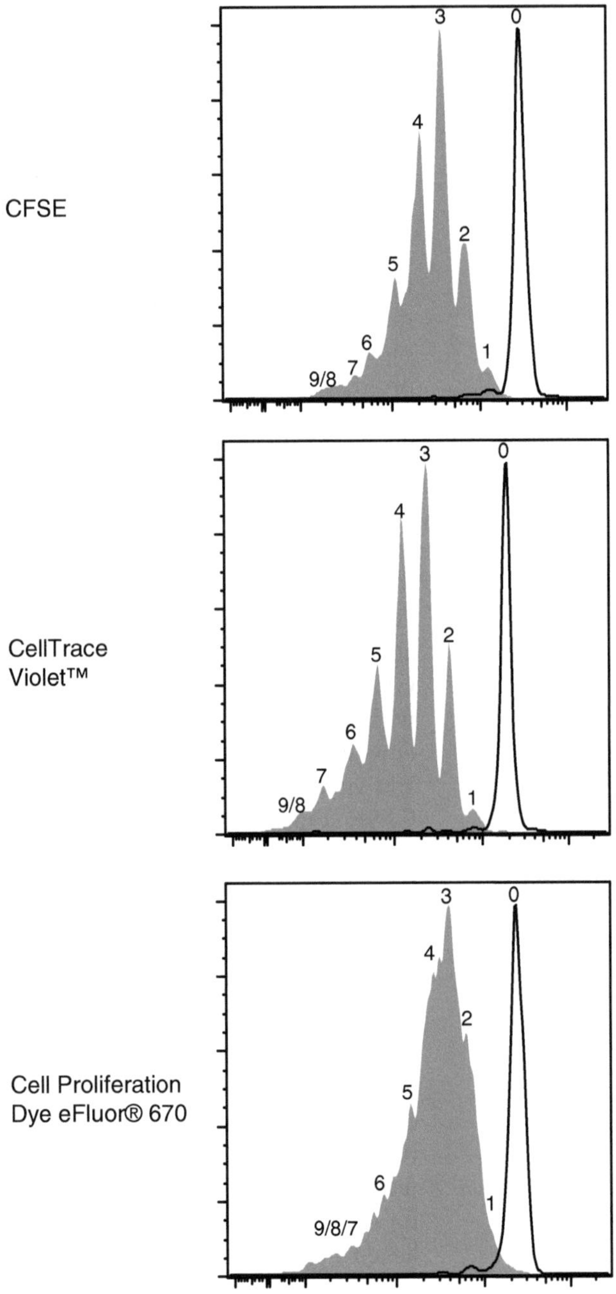

Fig. 1. Comparison of cell proliferation dyes. Polyclonal CD8⁺ T cells transferred to lymphoreplete mice (B6, open histograms) or lymphopenic mice with high levels of IL-7 (B6.129S7-*Il7r*^{tm1Imx}, shaded histograms). Whole lymphocytes from B6.PL-*Thy1*^{a} mice were labeled with CFSE or CellTrace Violet (CTV) or Cell Proliferation Dye eFluor 670 as described in Subheading 3. Labeled cells were adoptively transferred to separate hosts. After 7 days, host spleens were prepared as described in Subheading 3 and analyzed with a BD LSR-II flow cytometer. Histograms are gated on live, CD8^{+}CD90.1^{+} donor cells. Numbers above peaks mark generations 0 (undivided) through 8, and perhaps 9.

Table 1
Popular mouse models used in the study of homeostatic proliferation

Strain	Alternate names	Use	Notes	H2
C57BL/6	B6, Black 6	Host, donor	CD45.2$^{+/+}$ CD90.2$^{+/+}$	*b*
B6-SJL-*Ptprc*^{a} *Pepc*^{b}	Ly5a, Ly5.1	Host, donor	CD45.1$^{+/+}$ CD90.2$^{+/+}$	*b*
B6.PL-*Thy1*^{a}	Thy 1.1, Thy 1a	Host, donor	CD45.2$^{+/+}$ CD90.1$^{+/+}$	*b*
B6.12S7-*Rag1*$^{tm\ 1\ Mom}$	Rag 1$^{-/-}$	Host	Chronically lymphopenic	*b*
B6.12P2-*Tcrb*$^{tm\ 1\ Mom}$	TCRβ$^{-/-}$	Host	Chronically lymphopenic	*b*
C57BL/6-Tg(Tcra Tcrb)1100Mjb	OT-I	Donor	CD8 against ovalbumin	*b*
C57BL/6-Tg(Tcra Tcrb)425Cbn or C57BL/6-Tg(Tcra Tcrb)426-6Cbn[a]	OT-II	Donor	CD4 against ovalbumin	*b*
C57BL/6-Tg(TcrLCMV)318Sdz or C57BL/6-Tg(TcrLCMV)327Sdz[a]	P14	Donor	CD8 against LCMV	*b*
C57BL/6-Tg(TcrLCMV)Aox	SMARTA	Donor	CD4 against LCMV	*b*
C57BL/6-Tg(TcraH-Y, TcrbH-y) 1 Pas	Marilyn	Donor	CD4 against HY, a ubiquitous antigen	*b*
BALB/c	Balb	Host, donor	CD45.2$^{+/+}$ CD90.2$^{+/+}$	*d*
CBySmn.CB 17-*Prkdc*^{scid}	BALB/c *scid*	Host	Chronically lymphopenic	*d*
BALB/c- Tg(DO11.10) 10Dlo	DO11.10	Donor	CD4 against ovalbumin	*d*
Tg(TcrAND)53Hed	AND	Donor	B 10.BR or C57BL/6 background	*k,b*
B6.129P2-*Il2rb*$^{tm\ 1\ Mak}$	CD122$^{-/-}$	Host	Proliferation driven by IL-2 and IL-15	*b*
B6.129S7-*Il7r*$^{tm\ 1\ Imx}$	CD127$^{-/-}$	Host	High levels of IL-7	*b*
B6.129S4-*Il2rg*$^{tm\ 1\ Wjl}$	CD132$^{-/-}$	Host	Proliferation driven by IL-2, IL-7, and IL-15	*b*

[a]Two lines of mice exist for both the OT-II and P14 transgenes, according to Mouse Genome Informatics (http://www.informatics.jax.org)

1.5. Acutely Lymphopenic Hosts

Acutely lymphopenic hosts are generated by inducing a relatively short-term and transient depletion of lymphocytes from otherwise lymphoreplete wild type mice. Typical protocols to render wild type mice acutely lymphopenic, namely, by a sublethal dose of gamma irradiation or by injection of T-cell-specific monoclonal antibodies (mAbs), are described below. These approaches result in conditions that closely resemble iatrogenic lymphopenia caused by irradiation and/or cytotoxic drug treatment. Following the experimental lymphodepletion, residual host T cells also undergo homeostatic proliferation, but do not usually affect the behavior of the donor T cells because of the extreme severity of lymphopenia. Severe lymphopenia is maintained for 3–4 weeks after irradiation or anti-T mAb treatment. After this period, a rapid lymphoid reconstitution ensues from thymic production of new T cells, but this can be prevented by thymectomy prior to administering lymphodepleting treatment, rendering these mice lymphopenic almost indefinitely.

1.6. Chronically Lymphopenic Hosts

While there are many types of mice that exhibit chronic lymphopenia, for the sake of this discussion, we use the term *chronically lymphopenic* to refer specifically to hosts that are completely devoid of T cells due to a genetic defect (such as *Rag1*-deficient or *scid* mice). These are preferred hosts in instances when one is unable to render a wild type host lymphopenic. Furthermore, as discussed in the Subheading 3.1, using a host that lacks endogenous T cells can simplify the analysis by eliminating the need for a congenic marker on donor cells, as TCR expression will be unique to the donor cells (i.e., donor αβ T cells in $Rag1^{-/-}$ hosts are identifiable with fluorescent antibodies against TCRβ). However, when using chronically lymphopenic hosts, it is important to be aware that a fraction of polyclonal T cells will undergo explosive proliferation, presumably in response to antigens (Ags) of the commensal microbiota, which is not true homeostatic proliferation (28, 29). Thus, true homeostatic proliferation driven by self-components in chronically lymphopenic hosts needs to be studied with a mono- or oligoclonal population of T cells that does not cross-react with commensal microbial Ags.

1.7. Cytokine Receptor Deficient Mice

Cytokine receptor deficient mice are a class of hosts that has been less commonly used in studies of homeostatic proliferation. However, their use has advantages over other types of lymphopenic hosts. Mice that do not express receptors of the common γ-chain family (i.e., CD25, CD122, CD127, and CD132 itself) are unable to normally consume their cognate ligands, thus yielding increased levels of specific common γ-chain family cytokines (30–32). For instance, $CD127^{-/-}$ mice possess high concentrations of IL-7 and this can drive a slightly faster rate of homeostatic proliferation of $CD4^{+}$ and $CD8^{+}$ T cells than in $CD127^{+}$ lymphopenic hosts (3, 33, 34). More dramatically, an explosive rate of proliferation of T cells, especially $CD8^{+}$ cells, akin to normal antigen-driven proliferation, occurs in response to IL-2 and IL-15 in mice deficient in IL-2 receptor components ($CD122^{-/-}$ and $CD132^{-/-}$ hosts and to a lesser extent in $CD25^{-/-}$ mice) (31, 32). Since elevated levels of cytokines in these cytokine receptor-deficient hosts are due to lack of receptor expression, lymphodepleting treatment is not required in these mice for strong homeostatic proliferation of wild type donor T cells. This applies for both lymphopenic $CD127^{-/-}$ and $CD132^{-/-}$ hosts as well as $CD25^{-/-}$ and $CD122^{-/-}$ hosts that have lymphadenopathy.

New dye reagents and mouse models will facilitate experiments that probe the outstanding questions in the field of T cell homeostasis. For instance, homeostatic proliferation following irradiation and chemotherapy for cancer treatment may augment immune-mediated rejection of tumors, and therapies that capitalize on this opportunity could be very effective (35–37). Similarly, it remains to be determined whether exploitation of the known homeostatic factors can augment human vaccines (38, 39). In addition to research with direct clinical implications, we still have an incomplete understanding of some of the basic natural phenomena of T cell homeostasis. For instance, the rate of $CD4^+$ T cell homeostatic proliferation has long been observed to be slower than that of their $CD8^+$ counterparts; however, the reasons for and the implications of this difference are still incompletely understood (33, 34, 40).

Although the exact conditions will necessarily vary, the basic protocols are likely to remain the same for future experiments in homeostatic proliferation. The majority of experiments will continue to rely on the power of the mouse model. In order to isolate variables, researchers may find it necessary to generate bone marrow (BM) chimeras and to separate specific T cell populations from the lymphoid tissues. To observe homeostatic proliferation, it will be necessary to label donor cells with a proliferation dye and to introduce them into new, lymphopenic hosts. In the following sections, we present protocols one would find useful to study homeostatic proliferation. We have also made an effort to include background information, tricks, and data that would not be included in a typical publication covering homeostatic proliferation.

2. Materials

2.1. Bone Marrow Chimeras

1. Antibiotic water (0.08 mg/ml Trimethoprim and 0.4 mg/ml Sulfamethoxazole); 50× solution available from Hi·Tech Pharmacal Co. Inc.
2. Irradiator (e.g., Gammacell 40 Exactor from Best Theratronics, Ltd).
3. Surgical tools (scissors, forceps).
4. 60 mm polystyrene petri dish.
5. "Wash buffer"—high-glucose DMEM without l-glutamine, 0.5% heat-inactivated fetal calf serum (FCS), 10 mM HEPES.
6. 3 ml luer-lok syringe and 1 ml syringe.
7. 26 g and 28 g hypodermic needles.
8. Nylon cell filters.
9. Hemocytometer and microscope.

Table 2
Antibodies used in Subheadings 3.4 and 3.5

Antibody	Clone	Isotype	Sources tested	Final dilution[a]
Biotin-CD8	53–6.7	Rat IgG2a, κ	eBioscience	1:300
Biotin-CD4	RM4-5	Rat IgG2a, κ	BioLegend, eBioscience	1:300
Biotin-CD11b	M1/70	Rat IgG2b, κ	BioLegend, eBioscience	1:300
Biotin-CD11c	N418	Armenian Hamster IgG	BioLegend, eBioscience	1:300
Biotin-CD19	MB19-1	Mouse IgA, κ	BioLegend, eBioscience	1:300
Biotin-CD24	M1/69	Rat IgG2b, κ	BioLegend	1:500
Biotin-CD44	IM7	Rat IgG2b, κ	BioLegend, eBioscience	1:5,000
Biotin-CD45R/B220	RA3-6B2	Rat IgG2a, κ	BioLegend, eBioscience	1:500
Biotin-MHCII(I-A/I-E)	M5/114.15.2	Rat IgG2b, κ	BioLegend, eBioscience	1:700
Purified anti-CD16/32	2.4G2	Rat IgG2b, κ	BD Pharmingen[b]	1:75[b]

[a]Dilutions given are strictly for the enrichment protocol and will be excessive for cells at lower concentrations. The stock solution of all antibodies in this table is 0.5 mg/ml
[b]Not tested. This is an estimated dilution based on our experiences with in-house produced clone 2.4G2

10. Refrigerated centrifuge.
11. 14 ml snap cap, round bottom tubes (e.g., Becton Dickinson).
12. Broome-style mouse restraint.

2.2. Primary Lymphocyte Preparation from Lymph Nodes and Spleen

1. Pin board (e.g., styrofoam board) and 8 pins.
2. Surgical tools (scissors, forceps).
3. "Wash buffer"—see Subheading 2.1.
4. Dounce homogenizer with loose-fitting pestle.
   - or frosted microscope slides and 100 mm petri dish.
   - or 100 mm petri dish, two 1½″ 18 g needle, two 3 ml syringes.
   - or nylon cell filter, 3 ml syringe rubber plunger and 50 ml conical tube.
5. Refrigerated centrifuge.
6. Hemocytometer and microscope.

2.3. Negative Enrichment of T Cells

1. "Wash buffer"—see Subheading 2.1.
2. 14 ml or 5 ml snap cap, round bottom tubes.
3. Antibodies (see Table 2).
4. BD IMag™ Streptavidin Particles Plus-DM.
5. BD IMagnet™.

6. 50 ml conical tubes.
7. Glass Pasteur pipettes and rubber bulb (e.g., Fisher Scientific).
8. Refrigerated centrifuge.
9. Hemocytometer and microscope.

2.4. Labeling Cells with Cellular Dyes to Track Proliferation

1. "Wash buffer"—see Subheading 2.1.
2. "CFSE buffer"—1× PBS, 0.1% (w/v) Bovine Serum Albumin.
3. 14 ml snap cap, round bottom tube or 50 ml conical tubes (e.g., BD cat# 352070).
4. Nylon cell filters.
5. Cellular dye of your choice.
    - "CFSE" 5-(and 6-)carboxyfluorescein diacetate, succinimidyl ester (Invitrogen/Molecular Probes Cat. # C-1157 – MW 557.47).
    - "CTV" CellTrace™ Violet Cell Proliferation Kit for flow cytometry (Invitrogen/Molecular Probes cat# C34557).
    - "CPD670" Cell Prolifertion Dye eFluor® 670 (eBioscience cat. #65-0840-90 – MW 792.6).
6. Dimethyl sulfoxide (DMSO).
7. Refrigerated centrifuge.
8. Hemocytometer and microscope.

2.5. Generation of Acutely Lymphopenic Mice

2.5.1. Rendering Mice Lymphopenic by Gamma Ray Irradiation

1. Irradiator, for instance Gammacell 40 Exactor from Best Theratronics, Ltd.
2. Irradiator chamber for mice, also available from Best Theratronics, Ltd.

2.5.2. Rendering Mice Lymphopenic by Depletion of Thy1.2+ Cells

1. Anti-Thy1.2/CD90.2, clone 30-H12 (e.g., BioLegend).
2. 1× Phosphate Buffered Saline (PBS).
3. 28 g needles.
4. 1 ml syringes.

2.6. Adoptive Transfer of T Cells

1. 28 g needles.
2. 1 ml syringe.
3. Broome-style mouse restraint.
4. Gentle heat source.
5. 70% ethanol.

2.7. FACS Analysis

1. 5 ml polystyrene round-bottom tubes.
2. "FACS buffer" (1× PBS with 0.25% (v/v) heat-inactivated FCS and 0.05% (w/v) sodium azide).
3. Fluorescent molecule conjugated antibodies.
4. Propidium iodide.
5. Fc-blocking antibody, clone 2.4G2.

3. Methods

3.1. Considerations for Selecting Mice

The choice of appropriate host and donor mice is the single most important step in this protocol. An incorrect or hasty decision at this stage can lead to data that are difficult to interpret, or worse, data that confidently lead the researcher to false conclusions. It is important to consider (1) how to distinguish donor from host cells, (2) potential sources of strong antigenic or cytokine stimulation, (3) host and donor cells histocompatibility, and (4) methods to restrict genetic defects to specific cell types.

3.1.1. Distinguishing Donor from Host Cells

The simplest experiments involve transferring polyclonal T cells from one syngeneic mouse to another, with the donor cells distinguished by a congenic marker, such as Thy-1 (CD90) or Ly5 (CD45). For instance, B6.SJL-*Ptprc*^{a} *Pepc*^{b} or B6.PL-*Thy1*^{a} donor cells transferred to C57BL/6 (B6) hosts are easily identified during analysis by their expression of congenic markers CD45.1 or CD90.1, respectively (B6 mice express CD90.2 and CD45.2) (Table 1). When relying on the congenic marker CD90.1, it may be important to remember that mAbs to congenic CD90 proteins stain most, but not all T cells. We observe a very small population of CD90$^{int/lo}$ T cells with both anti-CD90.1 and anti-CD90.2 mAbs and with all conjugated fluorophores tested (Fig. 2 and data not shown). In this respect, CD45.1 (detected with monoclonal antibody A20) is expressed on all cells (Fig. 2) and hence is a better marker.

It is possible to conduct experiments without a congenic marker by relying on the fluorescence intensity above background of a cellular dye to identify the donor cells. Unfortunately, using this technique does not allow detection of donor cells that divide more than 6–8 times (when properly loaded with the dye). Cells that completely dilute the dye will be indistinguishable from the background fluorescence level of host cells and will be excluded from analysis. This method therefore is only useful for cells undergoing less than 6–8 cell divisions.

Alternatively, as discussed in the introduction, using host mice that are completely devoid of T cells alleviates the need for congenic markers as donor T cells can be identified by surface expression of TCR or CD3. However, it is important to remember the

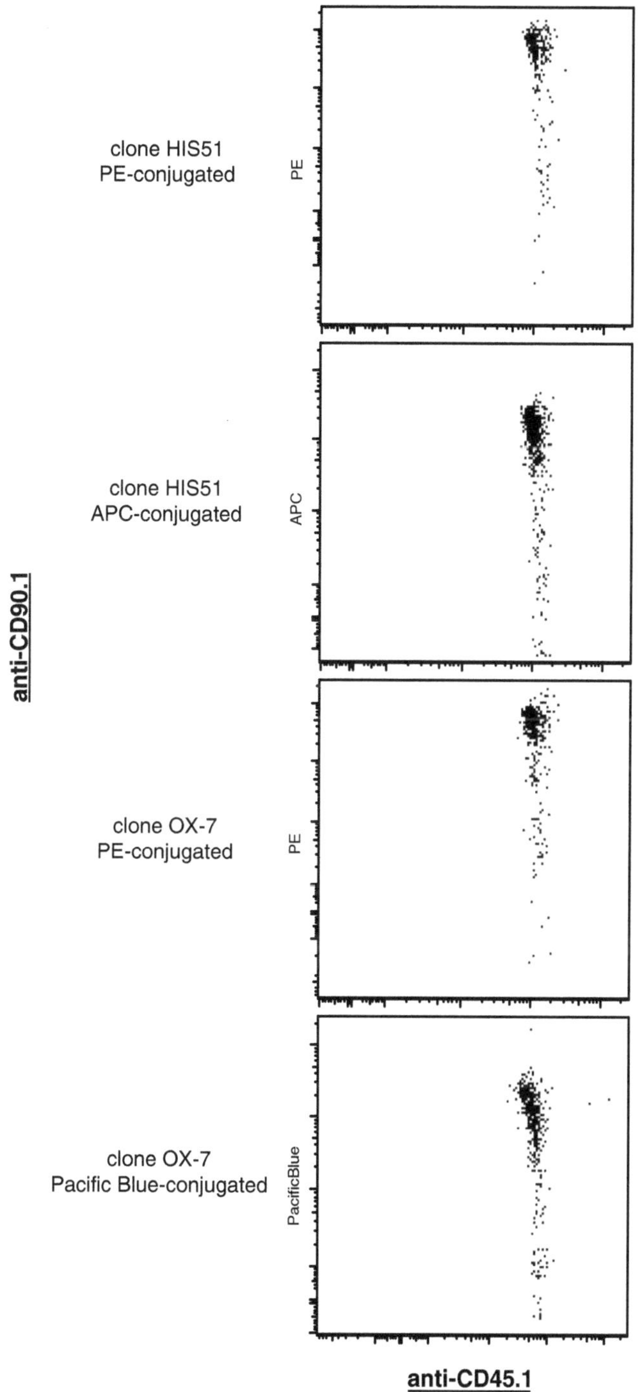

Fig. 2. Heterogeneous expression of CD90.1 contrasted with homogeneous expression of CD45.1 by $CD8^+$ T cells. $CD8^+$ lymphocytes purified from a donor mouse genetically homozygous for both CD90.1 and CD45.1 genes (B6.SJL-*Ptprc*^{a} *Pepc*^{b};B6.PL-*Thy1*^{a}) were CFSE labeled, transferred to a host B6 mouse and host LN prepared as described in methods after 1 week. Lymphocyte preparations were stained with antibodies against CD45.1, CD8 and one of an assortment of anti-CD90.1 antibodies as indicated in figure. Data were acquired on a BD LSR-II flow cytometer and the live $CFSE^+CD8^+$ cells are depicted.

caveat of using chronically lymphopenic hosts, as discussed below and more completely in Subheading 1.

3.1.2. Potential Sources of Strong Antigen or Cytokine Stimulation

Homeostatic proliferation is best observed as T cells undergo successive rounds of cellular division. However, not all T cell proliferation is necessarily homeostatic and the most common form of "proliferation contamination" is due to strong TCR stimulation. As discussed in the Introduction, it is not appropriate to transfer polyclonal donor T cells to chronically lymphopenic hosts (see Table 1), as a fraction of donor cells will mount a robust response to Ags of the gut (28). Similarly, to observe homeostatic proliferation when using TCR transgenic donor cells, it is essential to ensure that hosts are completely devoid of the cognate Ag (e.g., only virgin female hosts should be used with Marilyn TCR transgenic cells specific to an HY Ag).

Another form of proliferation contamination may come from intense cytokine stimulation as is seen in certain cytokine receptor deficient hosts (Table 1). For instance, while CD122$^{-/-}$ mice may seem attractive as hosts wherein to study the homeostatic cytokine IL-15, one must remember that these hosts will have increased levels of IL-2 as well as IL-15. Thus, the donor T cell proliferation observed in CD122$^{-/-}$ hosts is a result of stimulation from both cytokines.

3.1.3. Donor and Host Histocompatibility

As it may be necessary to transfer T cells between different strains of mice, it is important to consider the histocompatibility of those two strains. For example, when assessing the role of hypothetical "*Gene1*" in normal T cell homeostasis, one might transfer *Gene1*$^{-/-}$ cells to acutely lymphopenic B6 hosts. If the *Gene1*$^{-/-}$ mice are on a genetic background other than B6, histocompatibility issues may lead to an alloimmune response and the resulting data will not purely reflect a response to homeostatic factors. The MHC locus alleles (H2) are listed for the mouse strains in Table 1, but it is also important to be wary of minor histocompatibility mismatches. It is good practice to backcross mutant genes onto the genetic background of choice for 10 generations to remove minor mismatches. If it is uncertain whether or not host and donor mice are histocompatible, it may be necessary to perform a skin graft acceptance test (41).

3.1.4. Methods to Restrict Genetic Deficiencies to Specific Cell Types

Mice wherein every cell lacks or carries a certain gene are not always appropriate donor or host mice. For instance, it may be unclear whether the effect of a genetic deficiency is T cell intrinsic, or whether another cell type mediates the observed phenotype. In the absence of cell-specific conditional knockout mice, researchers may make chimeric mice with the bone marrow of one or more mouse genotypes.

3.2. Establishment of Bone Marrow Chimeras

To isolate the role of BM-derived cells (e.g., dendritic cells) from radio-resistant cells (e.g., fibroblastic reticular cells) in homeostatic proliferation, it is useful to generate BM chimeras. Here we describe a simple protocol to generate chimeric mice where the deficiency in hypothetical "*Gene2*" is limited to BM-derived cells.

1. Prior to obtaining BM, prepare the congenic wild type (CD45.1, *Gene2*$^{+/+}$) hosts. Place the mice on antibiotic feed or water on day -2.
2. On the following day (day -1), lethally irradiate the host mice with a dose of 1,000 cGy.
3. On day 0, euthanize donor CD45.2, *Gene2*$^{-/-}$ mice. Using dissecting tools, separate the skin from the limbs and then limbs from trunk. Harvest the femurs and tibias (see Note 1).
4. Remove the muscle tissue from the bones with scissors followed by earnest rubbing with a strong, dry paper towel (see Note 2). Place bones into a petri dish with 5–10 ml ice-cold wash buffer until you are ready to extract marrow.
5. Remove both ends of each long bone with scissors. Using a 3 ml syringe with a 26-gauge needle, pass cold wash buffer through the center of each bone to dislodge the marrow within.
6. Discard the empty bones and pass any clumps of bone marrow through the needle to make a single-cell suspension.
7. Transfer suspension into a 14 ml tube, bring to a known volume (pellet cells by ~300 × *g* at 8–10 °C and resuspend in cold wash buffer, if necessary) and determine the number of bone marrow cells therein with a hemocytometer. If necessary, deplete T cells from donor bone marrow at this stage. This may be accomplished by depletion with magnetic beads (i.e., follow Subheading 3.4 and use only biotinylated anti-CD4 and anti-CD8 antibodies in step 3) or by complement-mediated lysis (42).
8. Resuspend cells at a concentration of $2.5–25 \times 10^6$/ml (see Note 3) and inject 400 μl of this suspension into lethally irradiated *Gene2+/+* hosts (step 2) with a 1 ml syringe and 28-gauge needle via the lateral tail vein (as described in Subheading 3.8) or retro-orbital sinus.
9. Maintain mice on antibiotic water or feed for 2 weeks. Monitor the mice daily for signs of malaise (see Note 4).
10. Determine the efficiency of reconstitution by flow cytometric analysis of CD45 expression in blood cells (see Subheading 3.9 for a detailed staining protocol for flow cytometry). Successfully reconstituted bone marrow chimeric hosts will present with ~95% CD45.2^{+} CD45.1^{-} (donor) cells in the lymphocyte gate of a blood sample.

3.3. Primary Lymphocyte Preparation from Lymph Nodes

Here we describe a method to remove an assortment of lymph nodes (LN) that yields an average of 60×10^6 lymphocytes from a C57BL/6 mouse aged 2–6 months (see Note 5). We use the LN nomenclature described by Dunn (43), although an alternative nomenclature has been proposed more recently (44). Lymphocyte viability is adversely affected by adipose tissue that accompanies the harvested LN in the trituration step as well as time spent ex vivo. Thus, an effort should be made to separate adipose from lymphatic tissue and to return the cells to secondary hosts as soon as possible. LN of lymphopenic animals are quite small and we advise that one first become familiar with finding LN in wild type mice.

1. Euthanize donor mouse according to approved animal care and use protocols and arrange supine on pin board with the head at 12 o'clock and four feet pinned securely at 2, 10, 7, and 5 o'clock. Spray the mouse with 70% ethanol (see Note 6).
2. With forceps, pinch and raise skin in the area of the navel and make a 0.5–1 cm incision with scissors through the skin, but not into the peritoneal cavity. From this initial incision, perform three more cuts in the form of an inverted "Y" with points at the chin and ankles of the two hind limbs (see Note 7).
3. Using forceps, grasp the skin at points along the incision and pin them away from the mouse at roughly 8:30, 3:30, 10:30, and 1:30.
4. Remove the inguinal, axillary, brachial, cervical, and mesenteric LN and place in ice-cold wash buffer (see Note 8). If desired, remove the spleen as well (see Note 9).
5. Gently triturate LN or spleen (or both, separately) in cold wash buffer and then filter through 70–100 μm nylon mesh to prepare a single-cell suspension (see Note 10).
6. Pellet the cells at ~$300 \times g$ for 7 min at 8–10 °C and resuspend in a known volume of fresh, cold wash buffer (see Note 11).
7. Determine cell count by hemocytometer.
8. Proceed to Subheadings 3.4, 3.5 or 3.6 depending on your experimental requirements.

3.4. Negative Enrichment of $CD4^+$ T Cells (see Note 12)

1. Set aside a small aliquot (~10^5 cells) for downstream flow cytometry analysis.
2. Pellet the prepared cell suspension at ~$300 \times g$ for 7 min at 8 °C and resuspend the cells in cold wash buffer at a concentration of 1×10^8 cells/ml (see Note 13).
3. Add biotinylated (b-) antibodies: b-CD8, b-CD11b, b-CD11c, b-CD19, b-CD24, b-CD45R/B220 and purified anti-CD16/CD32 according to the final dilutions listed in (Table 2).

4. Incubate cells with antibodies for 25 min at 4 °C. To prevent sedimentation and improve the efficiency of antibody binding to cells, invert the tube 2–3 times during the incubation period.
5. Fill the tube to ~12 ml with cold wash buffer and centrifuge at ~300 × *g* for 7 min at 8 °C.
6. Resuspend the pellet in another 10 ml of cold wash buffer and centrifuge as in step 5 (see Note 14).
7. Resuspend the cell pellet in BD IMag™ Streptavidin Particles Plus-DM at 3 μl per 10^6 cells (see Note 15). Add an equivalent volume of cold wash buffer (i.e., 3 μl per 10^6 cells) and incubate on ice for 25 min. Gently vortex the tube 2–3 times during the incubation to mix and maintain suspension (see Note 16).
8. Add cold wash buffer to a volume of 9 ml.
9. Incubate tube in the IMagnet™ for 7 min at 4 °C. The streptavidin particles and bound cells will form a red/brown precipitate on the side of the tube that faces the magnet.
10. Using a glass Pasteur pipette, remove the wash buffer containing the unbound fraction to a new 50 ml conical tube while being careful not to disturb the red/brown precipitate.
11. Remove 14 ml tube from the magnet and resuspend the red/brown precipitate in 9 ml of cold wash buffer. Repeat steps 9–11 two additional times, combining the unbound fractions in a 50 ml conical tube.
12. Pellet the combined unbound fractions in the 50 ml conical tube at ~350 × *g* for 5 min at 8 °C. Resuspend the pellet in 5 ml of cold wash buffer and transfer to a new 14 ml snap cap tube. Collect any cells remaining in the empty 50 ml conical by rinsing with another 4 ml cold wash buffer and add to the 14 ml snap cap tube, bringing the volume to 9 ml.
13. Return the unbound fraction to the magnet, incubate for 7 min at 4 °C and, using a glass Pasteur pipette, carefully remove the unbound fraction (see Note 17).
14. Count the cells in the unbound fraction using a hemocytometer.
15. Verify the purity of the $CD4^+$ T cells via flow cytometric analysis of the whole lymphocyte aliquot (step 1), and a sample of the combined depleted fractions (step 14) (see Subheading 3.9). The combined unbound fractions should contain highly purified $CD4^+$ T cells.

3.5. Negative Enrichment of Naïve T Cells Using Magnetic Beads (see Note 18)

1. Set aside a small aliquot (~10^5 cells) for endpoint analysis of the initial memory ($CD44^{hi}$) T cell proportion.
2. Pellet the prepared cell suspension at ~350 × *g* for 7 min at 8 °C and resuspend the cells in cold wash buffer at a concentration of 10^8 cells/ml (see Note 13).

3. Add purified anti-CD16/CD32 antibodies to block Fc receptors and biotinylated (b-) antibodies specific for CD11b, CD11c, CD19, CD24, CD45R/B220, MHCII, and CD44 according to the working dilutions listed in Table 2. Also add a fluorophore-conjugated anti-mouse CD44 antibody (see Note 19).
4. Incubate cells with antibodies for 25 min at 4 °C. To prevent sedimentation and improve the efficiency of antibody binding to cells, invert the tube 2–3 times during the incubation period.
5. Wash twice with cold wash buffer as follows: Fill the test tube with cold wash buffer and pellet the cells at ~350 × *g* for 7 min at 8 °C. Resuspend the pellet in 1 ml wash buffer and then fill tube with cold wash buffer. Pellet the cells again by centrifugation (see Note 14).
6. Resuspend the pellet from in BD IMag™ Streptavidin Particles Plus-DM at 5 μl per 10^6 cells (see Note 15). Add an equivalent volume of cold wash buffer (i.e., 5 μl per 10^6 cells) and incubate on ice for 20–30 min.
7. Reconstitute the cells with cold wash buffer to a concentration of $0.5–1 \times 10^8$ cells/ml in round-bottom test tube (5 or 14 ml) and place the tube in the BD IMagnet™. Incubate 5 or 14 ml tubes at 4 °C in the magnet for 5 min or 8 min, respectively.
8. With the tube in the magnet, use a glass Pasteur pipette to carefully aspirate the supernatant (unbound fraction #1) and place in a fresh test tube.
9. Remove the tube from the magnet and resuspend the bound fraction (red/brown precipitate of cells, antibodies and magnetic particles) in cold wash buffer to the same volume as in step 7.
10. Place the tube in the BD IMagnet™ again. In this step, incubate 5 ml or 14 ml test tubes at 4 °C for 3 min or 5 min, respectively.
11. Using a new Pasteur pipette, carefully aspirate the supernatant (unbound fraction) and combine it with the unbound fraction from (step 8).
12. Verify the naïve phenotype of the T cells via flow cytometric analysis of the whole lymphocyte aliquot (see step 1), and a sample of the combined depleted fractions (see step 11) (see Subheading 3.9). The combined unbound fractions should contain highly purified $CD44^{lo}$ T cells as depicted in Fig. 3.
13. Pellet combined unbound fractions at ~350 × *g* for 7 min at 4 °C. Resuspend the cells in media or PBS and count using a hemocytometer. These cells are ready for downstream applications.

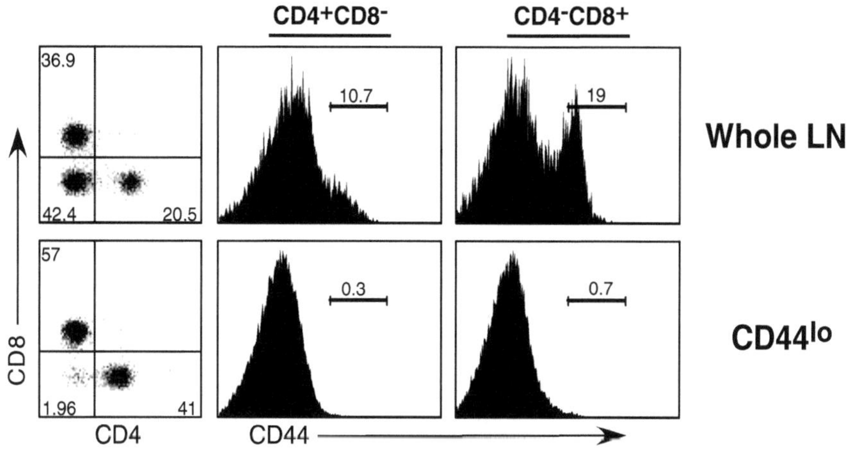

Fig. 3. Representative data of a successful enrichment of naïve $CD4^+$ and $CD8^+$ T cells. Non-T cells as well as $CD44^{hi}$ memory phenotype T cells were depleted from whole LN preparation (*top panels*) as described in Subheading 3.5 to yield a sample that is >97% $CD44^{lo}$ naïve T cells (*bottom panels*).

3.6. Labeling with Cellular Dyes to Track Proliferation

Here we describe our protocol for labeling cells with CFSE, CTV or CPD670 (see Note 20).

1. Prepare 5 mM solution of the dye (CFSE, CTV or CPD670) in DMSO (see Note 21). For experiments involving CFSE, prepare an aliquot of 0.1 mM CFSE in DMSO as well.
2. For labeling primary cells, gently triturate LN or spleen (see Subheading 3.3) in wash buffer to prepare a single cell suspension (see Note 22).
3. Determine number of cells to be labeled and, for each 10^7 cells, pre-warm 1 ml of CFSE buffer at 37 °C.
4. Pellet cells at ~300 × *g* for 7 min (in 14 ml tube) or 10 min (for 50 ml tube) at 8 °C. 14 ml tubes are appropriate for labeling $1–7 \times 10^7$ cells and 50 ml tubes should be used when labeling $7–25 \times 10^7$ cells.
5. Decant serum-containing media and resuspend the pellet in cold CFSE buffer before pelleting the cells again (see Note 23).
6. Thoroughly resuspend the cells to a final volume $1–2 \times 10^7$/ml in pre-warmed CFSE buffer (see Note 24).
7. As rapidly as possible, add 1 μl of 5 mM dye per ml of cell suspension and gently but thoroughly vortex. For CFSE labeling cells that will not be transferred in vivo (e.g., cells for in vitro culture or for FACS single-color controls), substitute 0.1 mM CFSE for 5 mM CFSE at this step.
8. Place tube in 37 °C water bath for 10 min.

9. To end the labeling reaction, fill the tube with cold wash buffer and pellet the cells at ~300 × *g* and for 7 min at 8 °C.
10. Resuspend cells in cold wash buffer and count them using a hemocytometer. One can expect a 20–30% loss of viable cells from the initial count. At this point, it is advisable to inject the labeled cells into host mice as soon as possible.

3.7. Generation of Acutely Lymphopenic Mice

The amplification of homeostatic proliferation observed in lymphopenic hosts makes them the ideal experimental models with which to study the phenomenon. Ablating genes essential to T cell development yields hosts with chronic lymphopenia (e.g., Rag-deficient, SCID, TCRβ$^{-/-}$, nude, etc.) and wild type mice may also be rendered acutely lymphopenic. Here we describe two methods for generating acutely lymphopenic mice: gamma-ray irradiation and host T cell-specific ablation by antibody. Irradiation is applicable to any mouse regardless of CD90 antigen; however, it relies on the availability of a gamma-ray source. Thus, we have included the alternative method for depleting host CD90.2^{+} cells with antibodies.

3.7.1. Rendering Mice Lymphopenic by Irradiation (see Note 25)

1. Determine the current dose rate of the radioactive source and calculate the duration of exposure to equal 6 Gy (i.e., 600 rads)
2. Secure mice in a vessel that is designed to fit with the irradiator (e.g., leucite pie chamber manufactured to the specifications of the Gammacell 40) and place vessel in the irradiator.
3. Close irradiator and ensure proper shielding is in place.
4. Administer 6 Gy to mice.
5. Once radiation has concluded and the source has safely returned to store position, open irradiator chamber and remove the vessel with the mice.
6. Return mice to housing cages and thoroughly clean and sterilize the vessel.

3.7.2. Rendering Mice Lymphopenic by Depletion of CD90.2 (Thy1.2)$^{+}$ Cells

1. Dilute purified anti-CD90.2^{+} (clone 30-H12) to 250 μg/ml in PBS (see Note 26).
2. Inject 400 μl (100 μg) of 30-H12 intraperitoneally (*i.p.*) on day -1 and day 0 of the adoptive transfer experiment. Donor cells are injected intravenously (*i.v.*) on day 0 and analysis is performed on day 7 (see Note 27).

3.8. Adoptive Transfer of Donor Lymphocytes into Lymphopenic Mice (see Note 28)

1. Each host mouse should be injected with a volume of 300–500 μl at a concentration <1.25 × 10^{8} cells/ml (see Note 29).
2. Immediately prior to injection, pass the cell solution through a nylon mesh filter (70–100 μm) to remove aggregates (see Note 30).

3. Attach a 28 gauge needle to a 1 ml syringe and load with the desired volume of cell suspension. Be careful to mitigate air bubbles in the syringe.
4. If available, use a gentle heat source (e.g., indirect heat lamp) to cause vasodilation of host mice.
5. Place host mouse in a Broome-style restraint with the tail exposed and locate the lateral tail vein. Avoid rupturing the ventral tail artery (see Note 31).
6. Introduce the needle into the lateral tail vein with the direction of blood flow (i.e., aimed toward the base of tail). Inject slowly, ensuring that the solution is entering *i.v.* and not subcutaneously (*s.c.*) (see Note 32).
7. Remove the needle and apply gentle pressure to the puncture site to stem blood loss and assist with wound closure.
8. Release the mouse from restraint and return it to the housing cage. Observe mice for 5 min to ensure that blood loss has ceased.

3.9. Analysis of In Vivo Homeostatic Proliferation

1. At a chosen timepoint between days 5–14 post-injection of labeled cells, collect and process the secondary lymphoid tissues (see Subheading 3.3) to create single cell suspensions (see Note 33). Determine the cell number from each tissue using a hemocytometer.
2. Aliquot a volume corresponding to $1–2 \times 10^6$ cells from each sample into 5 ml round bottom tubes. In addition to the experimental samples, aliquot 10^6 cells into tubes for single-color controls, one for each fluorescent antibody to be used in the analysis and one that will remain unstained.
3. Pellet cells at ~$400 \times g$ for 3 min at 8 °C.
4. Discard supernatants and resuspend the experimental samples in 100 μl FACS buffer containing appropriately diluted fluorescent molecule-conjugated antibodies and unconjugated Fc-blocking antibody 2.4 G2. Propidium iodide may also be included at this step at 1 μg/ml.
5. Resuspend single-color controls in 100 μl of FACS buffer alone (unstained control) or FACS buffer with a single fluorescent antibody or propidium iodide. Ideally, the single-color controls are stained with the same antibodies that are used to stain the experimental samples.
6. Incubate cells at 4 °C for at least 30 min.
7. Prepare a single-color control sample with the cellular dye used in the experiment (see Subheading 3.6). Remember to stain CFSE single-color controls at 0.1 μM rather than 5 μM CFSE.

8. Wash all samples with 1 ml cold FACS buffer and pellet at ~400 × g for 3 min at 8 °C.
9. Discard supernatants and resuspend the cells in 200 μl FACS buffer per tube. Samples are ready for analysis on a flow cytometer.

4. Notes

1. If necessary, bone marrow can be harvested from the long bones of both the hind and forelimbs but usually the femur and tibia will provide a sufficient quantity of bone marrow cells.
2. The most artful methods for removing muscle from bone are, unfortunately, clumsily transmitted through text. It is useful, but not essential to remove all muscle from bone.
3. One to two million bone marrow cells per mouse will usually be sufficient to rescue recipients from lethal irradiation.
4. Mice will remain healthy after 7 days if the donor bone marrow has grafted. However, mice remain immunocompromised for 4–5 weeks after transfer. Host mice are typically reconstituted with lymphocytes derived from donor BM cells 6 weeks after transfer.
5. Although T cells are available from both spleen and LN, the spleen yields a greater proportion of non-T cells than LN. Additionally, the spleen contains hematopoietic stem cells (HSCs) that can give rise to new mature T cells and disrupt the integrity of an experiment. Thus, LN are preferred instead of spleens as a source of T cells for adoptive transfer.
6. Dousing the mouse in ~10 ml 70% ethanol mitigates bacterial contamination and will control loose fur.
7. Blunt or rounded scissors are ideal for this procedure and may be used prior to the Y incisions to tunnel and separate skin from muscle.
8. LN are distinguished from adipose tissue as they are a shade toward beige while the adipose tissue is nearly white. The inguinal nodes are found on the flaps of skin created to the left and right of the mouse torso. If the inguinal is not immediately visible, it is in the fat pad at the junction of the three most prominent blood vessels in the flap. Next, find one axillary and 1-2 brachial nodes in the areas posterior to each forelimb. It is easiest to locate the axillary node before disturbing the tissue organization in this area. Once the axillary node has been removed, use forceps to expose the fat pad associated with the

triceps brachii, wherein the brachial node should be visible. Between the major salivary glands and the mandible, we generally find 2–4 superficial cervical nodes on either side of the mouse midline. These are distinguished from salivary glands by shape and texture as the nodes are rounded and will not tear as easily as the salivary glands. Furthermore, when ruptured, LN will appear to leak their contents and deflate while torn salivary glands retain their shape. The mesenteric nodes are harvested last and are easily located within the peritoneal cavity by first identifying the ceacum and then the ascending colon. The mesenteric nodes align to the mesenteric artery, parallel to the ascending colon and are difficult to separate completely from the associated adipose tissue.

9. If desired, the spleen is located posterior to the stomach and associated with the pancreas, which should be separated from the spleen as completely as possible.
10. The most efficient tool for triturating lymphoid tissue is a glass Dounce homogenizer that is tight, but has sufficient clearance between mortar and pestle so as not to shear cells. We generate these in house by wearing down new tight homogenizers with water and sand followed by smoothing with water alone. This is a lengthy process (7–10 h total), but may be worth the effort depending on how much tissue processing is foreseen in a lab. With this tool in hand, fill partially with cold wash buffer and squeeze LN or spleen gently between mortar and pestle until they are observed to "pop" and then gently plunge the pestle back and forth to wash the cells from within the tissue. A similar, but more cumbersome method involves a petri dish and two frosted microscope slides where the tissue is mechanically macerated between the frosted portions of the slides (which have previously been smoothed), using the Petri dish as a reservoir for the wash buffer and liberated cells. Alternatively, by assembling a 70–100 μm nylon mesh over a 50 ml conical tube, tissue may be pressed against the mesh with a rubber syringe plunger and intermittently passing cold wash buffer over the tissue through the mesh to rinse the liberated cells into the waiting conical tube. The most laborious, but perhaps most accessible and gentle method is executed with the tissue in a petri dish with cold wash buffer and using two 1½ in. 18 g needles bent at 90° angles and affixed to 3 ml syringes, one held to steady the tissue while the other is used to tear the node or spleen and then massage the cells out of the opening.
11. This step is important to remove the lipid from cellular debris or contaminating adipose tissue and should be executed soon after generating the single cell suspension, especially if the LN were small or accompanied by much adipose tissue.

12. We utilize a method to reliably enrich CD4+ T cells to >96% purity with negligible cellular loss from spleen or LN suspensions using BD IMag particles and commercially available antibodies. With minor adjustments this protocol may be used to enrich for CD8 T cells; however, in our hands CD8 purity is generally lower, ~90%. This protocol is a cost-effective alternative to the nearly identical commercially available kits.
13. Aliquot no more than 6 ml (i.e., 6×10^8 cells) per 14 ml tube. If necessary, divide sample into multiple 14 ml tubes
14. This step is important to ensure that any unbound antibody is removed from the cell suspension. Unbound antibody in the following steps will negatively affect purity.
15. Vortex magnetic particles before use.
16. When enriching CD4+ T cells from spleen, we advise that you use 5 μl streptavidin particles per 10^6 cells. Conversely, when enriching CD4+ T cells from LN, it is possible to use as little as 1 μl streptavidin particles per 10^6 cells, but you will observe slightly greater purity with 3 μl or 5 μl per 10^6 cells.
17. A red/brown precipitate may not be visible at this stage. This step ensures that any mag-bead-bound cells will be removed from the unbound fraction and increases the purity of the enriched sample.
18. Homeostatic proliferation causes naïve T cells to upregulate the markers that are associated with the memory T cell subset (i.e., $CD44^{hi}$ in C57BL/6 mice). Thus, in the endpoint analysis of an experiment in homeostatic proliferation it is impossible to determine which memory phenotype cells were naïve at the start of the experiment. In order to specifically analyze naïve T cell homeostatic proliferation, it is essential to remove the memory phenotype T cells from the whole tissue suspension. If a fluorescence-activated cell sorter is unavailable, it is possible to enrich for naïve T cells (i.e., $CD44^{lo}$ in C57BL/6 mice) by depleting the $CD44^{hi}$ memory subset with magnetic particles. This protocol typically results in naïve T cell populations with a purity of >98%. Outlined here are the steps involved in enriching for naïve T cells from C57BL/6 whole LN suspension (Fig. 3).
19. Prior to use, the two CD44 antibodies should be titrated to determine the concentrations at which they equivalently bind to lymphocytes. The addition of the fluorophore-conjugated antibody at this stage enables one to track the proportion of $CD44^{hi}$ T cells that are pulled off during the subsequent enrichment steps. FITC-CD44 was used in the example in Fig. 3.
20. The central technology in the study of homeostatic proliferation has been cellular labeling with CFSE. Alternative dyes have been developed, but none mark individual generations as clearly as CFSE or CTV from Invitrogen/Molecular Probes.

As discussed in the introduction, if you are unable to use CFSE or CTV, it is possible to use Cell Proliferation Dye eFluor® 670 (CPD670) from eBioscience.

21. We find it useful to store 5 μl aliquots of CFSE at −20 °C for future use.
22. For cells already in a single-cell suspension (i.e., cells from culture or sorted primary cells), it is not necessary to resuspend them in wash buffer; any pH buffered media with serum will suffice.
23. This wash step can be repeated if cells were previously in a high-serum media. Protein in the media, such as FCS, can interfere with labeling and it is important to transfer the cells to a low-protein media to optimize cell staining.
24. It is essential to use pre-warmed buffer to obtain uniform labeling, as efficiency of labeling is highly influenced by the temperature. If necessary, filter the cells through nylon cell filters to remove clumps of cellular detritus, which may form during the wash step. Large clumps may appear at this stage, but their size is misleading and does not necessarily correlate with cellular loss.
25. Ear tagging, notching, or other invasive procedures to identify host mice should be completed several days in advance, allowing the mouse to heal prior to lymphodepletion. Irradiation of host mice is conducted immediately prior to, or up to two days in advance of adoptive transfer of donor lymphocytes.
26. For experiments with donor cells that are $CD90.2^-$ (e.g., from a B6.PL-*Thy1*^{a} mouse) and hosts that are $CD90.2^+$ (e.g., C57BL/6), it is possible to specifically deplete host T cells with antibody clone 30-H12 (Rat IgG2b, kappa).
27. This dosing of 30-H12 is appropriate for analysis on days 5-8. Additional doses of 30-H12 may be required for longer experiments.
28. Given that T cells compete with each other for survival factors, it follows that injecting greater numbers of donor lymphocytes will decrease the rate and magnitude of homeostatic proliferation (45). Conversely, it is difficult to measure homeostatic proliferation when too few donor cells are injected (i.e., $<2 \times 10^5$ T cells). We typically inject 2×10^6 total lymphocytes or $0.5–1 \times 10^6$ purified $CD8^+$ or $CD4^+$ T cells via the lateral tail vein, described in Subheading 3.8, or retro-orbital sinus.
29. Smaller volumes increase the impact of injection error and larger volumes may be harmful to the host mouse. Prepare an excess of the cell suspension to facilitate syringe loading as even a slight touch to the sample tube will bend the needle tip and make it difficult to introduce into the vein.

30. Do not skip this step—aggregates of cellular debris may pose a lethal threat to host mice.

31. A quick rub with 70% ethanol will assist in visualization of lateral tail veins.

32. There will be much less plunger resistance when injecting *i.v.* versus *s.c.* Furthermore, a successful *i.v.* injection can be confirmed visually as the dark-colored lateral vein will rapidly lighten along the length of the tail.

33. In addition to the experimental mice, it is important to harvest LN or spleen cells from a mouse that did not receive dye-labeled cells; these cells will used as single-color controls when setting up the flow cytometer. LN are a superior to spleen as a source of cells for flow cytometric analysis in these experiments as they contain a greater proportion of T cells and fewer autofluorescent species.

References

1. Ernst B, Lee DS, Chang JM, Sprent J, Surh CD (1999) The peptide ligands mediating positive selection in the thymus control T cell survival and homeostatic proliferation in the periphery. Immunity 11:173–181
2. Murali-Krishna K, Ahmed R (2000) Cutting edge: naive T cells masquerading as memory cells. J Immunol 165:1733–1737
3. Schluns KS, Kieper WC, Jameson SC, Lefrancois L (2000) Interleukin-7 mediates the homeostasis of naive and memory CD8 T cells in vivo. Nat Immunol 1:426–432
4. Tan JT, Ernst B, Kieper WC, LeRoy E, Sprent J, Surh CD (2002) Interleukin (IL)-15 and IL-7 jointly regulate homeostatic proliferation of memory phenotype CD8+ cells but are not required for memory phenotype CD4+ cells. J Exp Med 195:1523–1532
5. Goldrath AW, Sivakumar PV, Glaccum M, Kennedy MK, Bevan MJ, Benoist C, Mathis D, Butz EA (2002) Cytokine requirements for acute and basal homeostatic proliferation of naive and memory CD8+ T cells. J Exp Med 195:1515–1522
6. Baccala R, Witherden D, Gonzalez-Quintial R, Dummer W, Surh CD, Havran WL, Theofilopoulos AN (2005) Gamma delta T cell homeostasis is controlled by IL-7 and IL-15 together with subset-specific factors. J Immunol 174:4606–4612
7. French JD, Roark CL, Born WK, O'Brien RL (2005) {gamma} {delta} T cell homeostasis is established in competition with {alpha} {beta} T cells and NK cells. Proc Natl Acad Sci U S A 102:14741–14746
8. Matsuda JL, Gapin L, Sidobre S, Kieper WC, Tan JT, Ceredig R, Surh CD, Kronenberg M (2002) Homeostasis of V alpha 14i NKT cells. Nat Immunol 3:966–974
9. Ranson T, Vosshenrich CA, Corcuff E, Richard O, Laloux V, Lehuen A, Di Santo JP (2003) IL-15 availability conditions homeostasis of peripheral natural killer T cells. Proc Natl Acad Sci U S A 100:2663–2668
10. Koka R, Burkett PR, Chien M, Chai S, Chan F, Lodolce JP, Boone DL, Ma A (2003) Interleukin (IL)-15R[alpha]-deficient natural killer cells survive in normal but not IL-15R[alpha]-deficient mice. J Exp Med 197:977–984
11. Prlic M, Blazar BR, Farrar MA, Jameson SC (2003) In vivo survival and homeostatic proliferation of natural killer cells. J Exp Med 197:967–976
12. Ranson T, Vosshenrich CA, Corcuff E, Richard O, Muller W, Di Santo JP (2003) IL-15 is an essential mediator of peripheral NK-cell homeostasis. Blood 101:4887–4893
13. Jamieson AM, Isnard P, Dorfman JR, Coles MC, Raulet DH (2004) Turnover and proliferation of NK cells in steady state and lymphopenic conditions. J Immunol 172:864–870
14. Purton JF, Tan JT, Rubinstein MP, Kim DM, Sprent J, Surh CD (2007) Antiviral CD4+ memory T cells are IL-15 dependent. J Exp Med 204:951–961
15. Sprent J, Surh CD (2011) Normal T cell homeostasis: the conversion of naive cells into memory-phenotype cells. Nat Immunol 12: 478–484

16. Michalek RD, Rathmell JC (2010) The metabolic life and times of a T-cell. Immunol Rev 236:190–202
17. Osborne LC, Abraham N (2010) Regulation of memory T cells by gammac cytokines. Cytokine 50:105–113
18. Takada K, Jameson SC (2009) Naive T cell homeostasis: from awareness of space to a sense of place. Nat Rev Immunol 9:823–832
19. Surh CD, Sprent J (2008) Homeostasis of naive and memory T cells. Immunity 29:848–862
20. Rochman Y, Spolski R, Leonard WJ (2009) New insights into the regulation of T cells by gamma(c) family cytokines. Nat Rev Immunol 9:480–490
21. van Leeuwen EM, Sprent J, Surh CD (2009) Generation and maintenance of memory CD4(+) T Cells. Curr Opin Immunol 21:167–172
22. Parish CR (1999) Fluorescent dyes for lymphocyte migration and proliferation studies. Immunol Cell Biol 77:499–508
23. Shapiro HM, John Wiley and Sons (2003) Practical flow cytometry. Wiley-Liss, New York, pp 371–374
24. Bantly AD, Gray BD, Breslin E, Weinstein EG, Muirhead KA, Ohlsson-Wilhelm BM, Moore JS (2007) Cell Vue Claret, a new far-red dye, facilitates polychromatic assessment of immune cell proliferation. Immunol Invest 36: 581–605
25. Yen MH, Lepak N, Swain SL (2002) Induction of CD4 T cell changes in murine AIDS is dependent on costimulation and involves a dysregulation of homeostasis. J Immunol 169: 722–731
26. Hawkins ED, Hommel M, Turner ML, Battye FL, Markham JF, Hodgkin PD (2007) Measuring lymphocyte proliferation, survival and differentiation using CFSE time-series data. Nat Protoc 2:2057–2067
27. Quah BJ, Warren HS, Parish CR (2007) Monitoring lymphocyte proliferation in vitro and in vivo with the intracellular fluorescent dye carboxyfluorescein diacetate succinimidyl ester. Nat Protoc 2:2049–2056
28. Kieper WC, Troy A, Burghardt JT, Ramsey C, Lee JY, Jiang HQ, Dummer W, Shen H, Cebra JJ, Surh CD (2005) Recent immune status determines the source of antigens that drive homeostatic T cell expansion. J Immunol 174:3158–3163
29. Min B, Yamane H, Hu-Li J, Paul WE (2005) Spontaneous and homeostatic proliferation of CD4 T cells are regulated by different mechanisms. J Immunol 174:6039–6044
30. Tsunobuchi H, Nishimura H, Goshima F, Daikoku T, Nishiyama Y, Yoshikai Y (2000) Memory-type CD8+ T cells protect IL-2 receptor alpha-deficient mice from systemic infection with herpes simplex virus type 2. J Immunol 165:4552–4560
31. Cho JH, Boyman O, Kim HO, Hahm B, Rubinstein MP, Ramsey C, Kim DM, Surh CD, Sprent J (2007) An intense form of homeostatic proliferation of naive CD8+ cells driven by IL-2. J Exp Med 204:1787–1801
32. Ramsey C, Rubinstein MP, Kim DM, Cho JH, Sprent J, Surh CD (2008) The lymphopenic environment of CD132 (common gamma-chain)-deficient hosts elicits rapid homeostatic proliferation of naive T cells via IL-15. J Immunol 180:5320–5326
33. Guimond M, Veenstra RG, Grindler DJ, Zhang H, Cui Y, Murphy RD, Kim SY, Na R, Hennighausen L, Kurtulus S, Erman B, Matzinger P, Merchant MS, Mackall CL (2009) Interleukin 7 signaling in dendritic cells regulates the homeostatic proliferation and niche size of CD4+ T cells. Nat Immunol 10:149–157
34. Martin CE, Kim DM, Sprent J, Surh CD (2010) Is IL-7 from dendritic cells essential for the homeostasis of CD4+ T cells? Nat Immunol 11:547–548, author reply 8
35. Asavaroengchai W, Kotera Y, Mule JJ (2002) Tumor lysate-pulsed dendritic cells can elicit an effective antitumor immune response during early lymphoid recovery. Proc Natl Acad Sci U S A 99:931–936
36. Dudley ME, Wunderlich JR, Robbins PF, Yang JC, Hwu P, Schwartzentruber DJ, Topalian SL, Sherry R, Restifo NP, Hubicki AM, Robinson MR, Raffeld M, Duray P, Seipp CA, Rogers-Freezer L, Morton KE, Mavroukakis SA, White DE, Rosenberg SA (2002) Cancer regression and autoimmunity in patients after clonal repopulation with antitumor lymphocytes. Science 298:850–854
37. Dummer W, Niethammer AG, Baccala R, Lawson BR, Wagner N, Reisfeld RA, Theofilopoulos AN (2002) T cell homeostatic proliferation elicits effective antitumor autoimmunity. J Clin Invest 110:185–192
38. Melchionda F, Fry TJ, Milliron MJ, McKirdy MA, Tagaya Y, Mackall CL (2005) Adjuvant IL-7 or IL-15 overcomes immunodominance and improves survival of the CD8+ memory cell pool. J Clin Invest 115:1177–1187
39. Pellegrini M, Calzascia T, Elford AR, Shahinian A, Lin AE, Dissanayake D, Dhanji S, Nguyen LT, Gronski MA, Morre M, Assouline B, Lahl K, Sparwasser T, Ohashi PS, Mak TW (2009) Adjuvant IL-7 antagonizes multiple cellular

and molecular inhibitory networks to enhance immunotherapies. Nat Med 15:528–536

40. Ferreira C, Barthlott T, Garcia S, Zamoyska R, Stockinger B (2000) Differential survival of naive CD4 and CD8 T cells. J Immunol 165:3689–3694
41. McFarland HI, Rosenberg AS (2009) Skin allograft rejection. In: John E. Coligan et al. (eds) Current protocols in immunology. Chapter 4: Unit 4
42. Hathcock KS (2001) T cell depletion by cytotoxic elimination. Curr Protoc Immunol. Chapter 3: Unit 3 4
43. Dunn TB (1954) Normal and pathologic anatomy of the reticular tissue in laboratory mice, with a classification and discussion of neoplasms. J Natl Cancer Inst 14:1281–1433
44. Van den Broeck W, Derore A, Simoens P (2006) Anatomy and nomenclature of murine lymph nodes: descriptive study and nomenclatory standardization in BALB/cAnNCrl mice. J Immunol Methods 312:12–19
45. Dummer W, Ernst B, LeRoy E, Lee D, Surh C (2001) Autologous regulation of naive T cell homeostasis within the T cell compartment. J Immunol 166:2460–2468

Chapter 10

Quantitating Lymphocyte Homeostasis In Vivo in Humans Using Stable Isotope Tracers

Liset Westera, Yan Zhang, Kiki Tesselaar, José A.M. Borghans, and Derek C. Macallan

Abstract

Humans have a remarkable ability to maintain relatively constant lymphocyte numbers across many decades, from puberty to old-age, despite a multitude of infectious and other challenges and a dramatic decline in thymic output. This phenomenon, lymphocyte homeostasis, is achieved by matching the production, death, and phenotype transition rates across a network of varied lymphocyte subpopulations. Understanding this process in humans depends on the ability to measure in vivo rates of lymphocyte production and loss. Such investigations have been greatly facilitated by the advent of stable isotope labeling approaches, which use the rate of incorporation of a tracer into cellular DNA as a marker of cell division. Two labeling approaches are commonly employed, one using deuterium-labeled glucose and the other using deuterium-labeled water, also known as heavy water ($^{2}H_{2}O$). Here we describe the application of these two labeling techniques for measurement of human in vivo lymphocyte kinetics through the four phases of investigation: labeling, sampling, analysis, and interpretation.

Key words: Lymphocyte, Homeostasis, Memory, Naive, Kinetics, Turnover, Dynamics, Isotope, Tracer, Heavy water, Deuterated glucose, Proliferation, Production, Cell death, Labeling

1. Introduction

1.1. Quantification of Lymphocyte Dynamics in Human Clinical Studies

Long-term maintenance of immune homeostasis is essential for proper functioning of the immune system, as illustrated by several diseases including HIV infection and leukemia, in which immune homeostasis is disturbed. Immune homeostasis is directly linked to the dynamics of the lymphocytes that comprise the cellular immune system. In order for populations to be maintained over time, cell production and loss must be balanced. To better understand normal lymphocyte homeostasis and the pathology of conditions that perturb lymphocyte numbers, quantification of production and loss rates of lymphocytes is crucial (1).

Andrew L. Snow and Michael J. Lenardo (eds.), *Immune Homeostasis: Methods and Protocols*, Methods in Molecular Biology, vol. 979, DOI 10.1007/978-1-62703-290-2_10, © Springer Science+Business Media New York 2013

Several approaches have been used to quantify lymphocyte dynamics, including measurement of telomere length or Ki-67 expression, quantitation of T-cell receptor excision circles (TRECs), tracking cells with radiation-induced chromosomal damage, cell labeling with carboxyfluorescin diacetate succinimidyl ester (CFSE), and DNA labeling with the thymidine analog 5′-bromo-3′-deoxyuridine (BrdU). Although all of these approaches have contributed significantly to our understanding of lymphocyte dynamics, most are not applicable to in vivo human clinical studies. By contrast, recently introduced stable isotope labeling techniques have made the in vivo investigation of lymphocyte dynamics in humans feasible (2–7). Stable isotopes have the great advantage of being suitable for human (clinical) investigation since they are nonradioactive and have no inherent toxicity in tracer doses.

1.2. Stable Isotope Labeling

The key to the use of stable isotopes to measure cell turnover is the link between cell division and DNA synthesis (2, 4). Although DNA synthesis occurs in S-phase of the cell cycle, it can be used as a surrogate marker for proliferation at the population level (3). Similarly, since non-replicative DNA turnover (including non-replicative DNA repair, nucleoside substitutions during RNA–DNA interactions during message transcription and DNA unfolding) is relatively slow, loss of labeled DNA can be used as a marker of death of labeled cells, or of disappearance of labeled cells from the compartment being sampled.

Stable isotope tracer studies are based on a generic principle; when a product arises from a labeled precursor, the rate of incorporation of label from precursor to product yields the synthesis rate of the product (8). In this case the product is DNA and the immediate precursor is deoxynucleotide triphosphate (dNTP) (2, 4, 5). The precursor pool can be labeled in in vivo studies using deuterium-labeled glucose or water. Both result in deuterium-labeled dNTP through the de novo nucleotide synthesis (DNNS) pathway (2). Glucose carries two deuterium atoms into the pentose moieties, which form the building blocks for nucleotide synthesis; the label is thus specifically incorporated into the sugar-phosphate backbone of DNA. Deuterated water-derived deuterium is incorporated at multiple sites (C–H bonds) within the newly synthesized purine dNTPs in DNA (2, 3, 6). Hence, cells that replicate in the presence of deuterium-labeled precursors, either glucose or water, will incorporate deuterium into the DNA of their progeny, whilst non-dividing cells remain unlabeled. The label is retained within the progeny, even if the cell divides again, until the cell dies or leaves the pool of cells under investigation, either by localization elsewhere or by phenotype transition (3).

One advantage of the use of deuterated glucose or heavy water over thymidine nucleoside analogue labels such as BrdU is that deuterated glucose and heavy water predominantly label through the *de novo* nucleoside synthesis pathway, whereas thymidine nucleoside analogues enter cellular dNTP pools via the salvage pathways. Salvage pathway activity is highly variable, being modulated by cellular physiology and this variability may interfere with BrdU labeling in an unpredictable way (9, 10). Quantitation is achieved by analyzing (by mass spectrometry) the fraction of deuterium-labeled moieties within the DNA of sampled cell populations during and/or after labeling. Analysis only allows conclusions to be drawn about populations of cells. By contrast, approaches such as BrdU labeling, ^{3}H-thymidine labeling, Ki-67 expression analysis or CFSE dilution enable cell-by-cell analysis, which should be seen as yielding complementary data. A population of cells with a high rate of turnover will incorporate large amounts of the isotope deuterium, whereas one with few mitoses will incorporate little.

In this paper we describe the use of deuterium-labeled compounds to quantify production and loss rates of different types of lymphocytes and demonstrate how mathematical modeling may be used to interpret the experimental data (11–14). Although stable isotope labeling has many potential applications (virtually any cell type can be studied) (1), this protocol is limited to its application to the study of human lymphocyte homeostasis. Modification for animal studies is also possible.

The next section introduces the four different components that characterize a stable isotope labeling experiment from beginning to end: (1) labeling, (2) sampling, (3) analysis, and (4) interpretation (Fig. 1.).

1.3. The Four Components of a Labeling Experiment

1.3.1. Labeling

Glucose or Water?

The expected turnover rate of a cell population is a key determinant in the choice between labeling with deuterated water or glucose. Because of its rapid "on" and "off" kinetics, glucose is appropriate for cells with rapid turnover. Conversely, deuterated water is the method of choice when analyzing slow turnover populations since it can be administered over long periods of time (even months if necessary). The resulting long-lasting enrichment in body water permits a reliable labeling of cell populations that rarely divide, such as human naive T cells. Importantly, it is impossible to use heavy water labeling for the study of cell types that turn over much more rapidly than water itself. For such cell types, glucose labeling is the appropriate method.

Deuterated Glucose Labeling

Glucose may be given either orally or by intravenous infusion. Oral labeling is less invasive but requires at least half-hourly administration and for this reason overnight administration is difficult. Oral labeling for 10 h gave sufficient signal to measure the kinetics of $CD4^{+}CD45RO^{+}CD25^{hi}$ T cells, but not slowly-dividing $CD4^{+}CD45RA^{+}$ T cells, in which enrichment was not quantifiable (15). Intravenous infusion allows a longer labeling phase but is more invasive and requires more intensive attention to safety issues

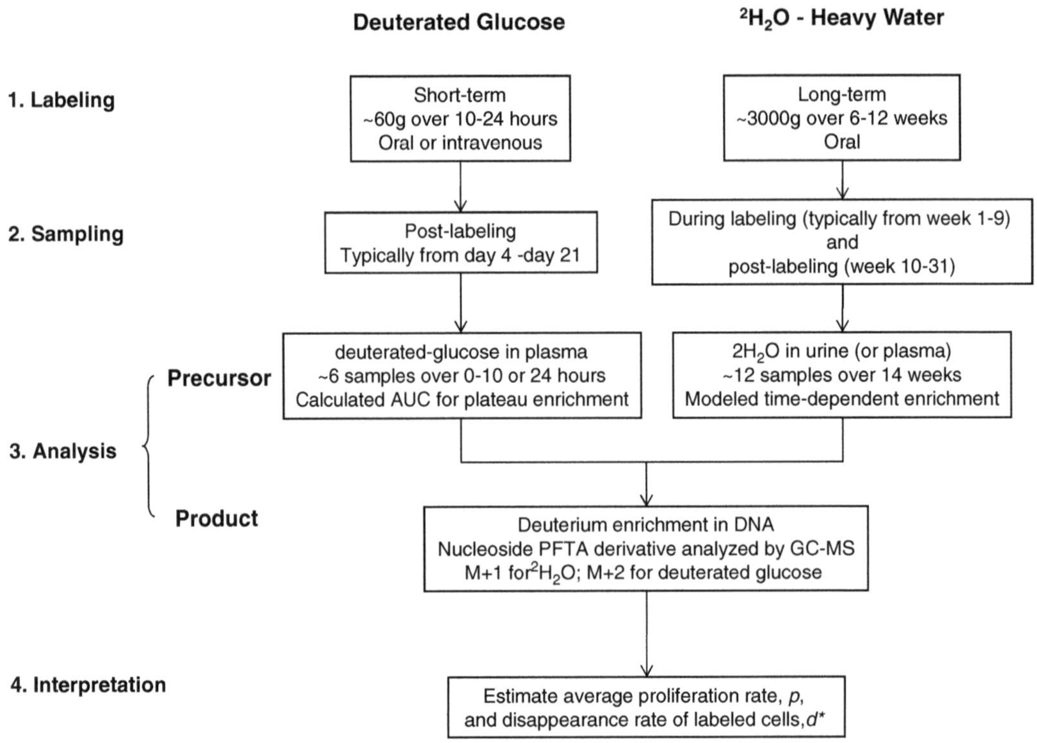

Fig. 1. **Overview of stable isotope labeling strategies**. Comparison of deuterated glucose and heavy water labeling approaches.

such as the sterility/pyrogenicity of the infusate. Labeling for 24 h is adequate to capture the kinetics of CD45RO$^+$ and CD45RA$^+$ T cells,(12, 16–18) B cells (19), central and effector memory T cells (17), regulatory T cells (7, 15), and leukemic cells in chronic lymphocytic leukemia (20). Longer infusions have been used, such as 48 h (21–23) or 5 days (24–26). The sustained levels of enrichment achieved by longer infusions allow analysis of less-rapidly dividing cell populations (5), but for very slowly dividing populations, the heavy water approach should be preferred.

Heavy Water Labeling

Heavy water labeling experiments in humans typically start with an oral bolus (or ramp-up) to achieve near-plateau body water enrichment, after which smaller daily doses are given (orally) to maintain body water enrichment at approximately the same level over a timescale of several weeks. During intake of the initial bolus subjects may experience transient vertigo or nausea due to an effect on the inner ear vestibular apparatus (see Note 8) (27). Both the level of body water enrichment that is achieved and the duration of exposure to enriched body water influence the extent of label incorporation by a given cell population; depending on the cell population of interest, both the intended body water enrichment and the duration of labeling can be varied. Labeling protocols to study lymphocyte dynamics typically reach long-term body water enrichments of 1–3% (6, 13, 22, 28, 29). Heavy water administration can in principle be continued as long as necessary for the cell populations under investigation; for lymphocyte studies the labeling period has typically been in the order of a few months (6, 13, 22, 28, 29). Continuous administration for 9 weeks has been successfully applied to measure lymphocyte turnover in healthy individuals (13), in B-cell chronic lymphocytic leukemia (B-CLL) patients (28), and in HIV-1 patients (Vrisekoop, unpublished), and resulted in readily measurable label incorporation even by very slowly dividing human naive ($CD27^{+}CD45RO^{-}$) T cells (13).

1.3.2. Sampling

There are two elements to sampling: timing and cell selection. (A further consideration is the "compartment" in which the cells reside, but peripheral blood is the only compartment considered herein.) In both labeling approaches, both the precursor and product are sampled.

Deuterated Glucose Labeling

For glucose labeling, glucose enrichment in plasma (precursor) must be measured at sufficient time-points during the labeling phase to derive an area-under-curve estimate of enrichment × time (see Note 13) (5). The number of samples is a compromise between reducing errors of estimation and reducing subject discomfort. Typically we use 6 time-points for 10-h oral labeling. Pinprick samples yield more than sufficient sample, although an indwelling intravenous catheter is an alternative (Fig. 2.).

For DNA analysis (product), all samples are typically taken during the post-labeling phase, although longer infusion studies have also collected up-labeling data (24–26). We assume that maximal intracellular labeling occurs at or very shortly after the end of the glucose administration phase on the basis that blood glucose and intracellular dNTP pools are likely to be small and short-lived; $t½$ for blood glucose is <2 h and in vitro DNA labeling occurs within an hour of introduction of labeled glucose to the culture media (Macallan, unpublished observations). However, maximal labeling of circulating cells in blood does not occur until considerably later; the length of this "lag" phase has not been well-defined but we find what appears to be a delay of about 2–3 days between division, presumably in lymphoid compartments or tissues, and release into the bloodstream (5). We therefore suggest delaying initial sampling until ≥ day 3. After this peak of labeling, multiple follow-up samples are collected from which the progressive loss of label can be monitored.

Heavy Water Labeling

For heavy water labeling, body water $^{2}H_2O$ enrichment (precursor) must be well monitored both during the labeling and post-labeling phases. Because it takes several days to reach plateau levels of heavy water in the body fluids (plasma, urine), it is essential to sample frequently in the period from start of labeling until steady-state levels

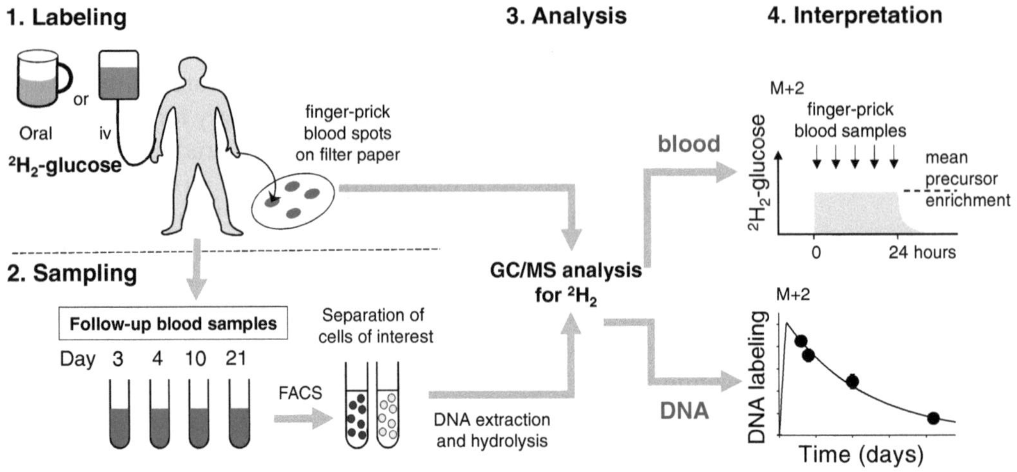

Fig. 2. **General schematic of protocol for analysis of lymphocyte kinetics using deuterium-labeled glucose.** The example shown illustrates how either oral or intravenous administration of (6,6-$^2H_2$)-glucose may be used to label dividing lymphocytes in vivo. The deuterium content of DNA, analyzed by gas chromatography-mass spectrometry (GC/MS) is compared to the average glucose enrichment in plasma, E_{Glu} (both M+2) to derive fractional labeling curves over time. FACS, fluorescence activated cell sorting. Modified from Macallan et al. (5).

of heavy water have been reached. Similarly, heavy water has a slow wash-out after labeling has been terminated, hence monitoring how body water enrichment decreases post-labeling is important. Plasma or urine can be used, but urine is the most practical because urine samples are easy for subjects to collect at home. Enough samples can thus readily be collected for reliable curve-fitting (Fig. 3).

For DNA analysis (product), samples are taken during the labeling and post-labeling phases. Cell analysis can be less frequent than urine sampling and is a compromise between reducing errors of estimation and reducing subject discomfort. Since a great deal of information can be obtained from the start of the early labeling phase and the start of the post-labeling phase, it is important to collect sufficient early sampling points. Typically, 6 samples (in addition to the baseline sample) are taken during the labeling phase, and 7 samples post-labeling, but this can be varied according to the cell type investigated and the subjects that are included.

The maximum level of label incorporation that cells can possibly obtain should be determined by including granulocytes (or another cell type with rapid turnover, such as monocytes), in the analysis. Label incorporation by cell populations of interest can be scaled between 0% and 100% by normalizing to this granulocyte maximum.

Details of how to sort the cells of interest are beyond the scope of this article. However, cell population purity needs to be considered carefully. Interpretation artefacts are likely if a high turnover population contaminates a low turnover one; hence for example, excluding monocytes (high turnover) from lymphocyte preparations is critical.

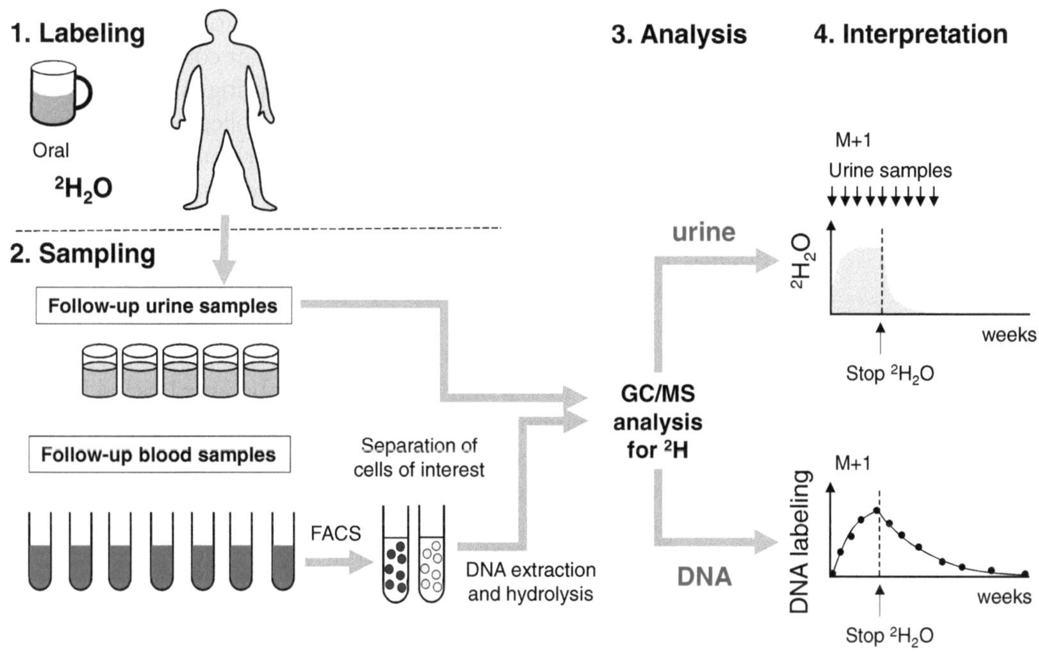

Fig. 3. **General schematic (as Fig. 2.) of protocol for analysis of lymphocyte kinetics using deuterium-labeled water.** The example shown illustrates how oral intake of $2H_2O$ may be used to label dividing lymphocytes in vivo. In addition to measuring the deuterium content in DNA, urine samples are taken to allow monitoring of the deuterium enrichment in body water over time. Both DNA and urine deuterium enrichment are analyzed for M+1 enrichment by gas chromatography-mass spectrometry (GC/MS).

1.3.3. Analysis

Isotopic Measurement

Isotopic enrichments are measured by mass spectrometry. Several approaches are possible. We use gas chromatography/mass spectrometry (GC/MS), which is sensitive, reproducible, and requires a relatively modest hardware investment. Alternatively, LC/MS approaches (30) require less sample processing whilst gas chromatography/ pyrolysis/ isotope ratio–mass spectrometry is able to measure very low enrichments but requires larger cell numbers (>10^7) (31). For GC/MS, derivatization is required and most analyses of DNA enrichment have either used a pentose tetra-acetate (PTA) derivative, analyzed by positive chemical ionization (PCI) (32) or a pentafluoro tri-acetate (PFTA) derivative which ionizes well under negative chemical ionization (NCI). The latter approach was described by Busch et al. (2) and is reproduced here with minor modifications. The same analysis is used for glucose and water labeling except that glucose labeling is monitored by measuring the ion with two additional mass units (M + 2) and water labeling is monitored by measuring the ion with one additional mass unit (M + 1) (2). For analysis of (precursor) deuterium enrichment in glucose and urine, several alternative approaches have been described. We describe the procedures used in our laboratories.

1.3.4. Interpretation

The use of appropriate mathematical models to translate the enrichment data into biologically meaningful parameters is a critical step in this type of investigation. This is not straightforward and several approaches have been described. One important consideration is the discrepancy between information collected during labeling (typically for heavy water studies) and information collected post-labeling (both for deuterated glucose and heavy water studies). The rate of label acquisition during labeling is determined by the average turnover rate of the population: all cells, i.e., cells that divided and incorporated label and non-divided unlabeled cells, together determine the fraction of label in the total cell population. This part of the labeling curve therefore reflects what is happening in the population as a whole. In contrast, the rate of loss of label during the post-labeling phase reflects the loss of only those cells that became labeled because of recent cell division. If cell populations are kinetically heterogeneous, i.e., if they consist of 2 or more sub-populations with different turnover characteristics, the loss rate of labeled cells tends to be higher than the average turnover rate of the population as a whole, because the labeled fraction is biased towards cells with rapid turnover (1, 11). Only if populations are kinetically homogeneous, i.e., if all cells have the same rate of turnover, will the loss rate of labeled cells truly reflect the average turnover rate. Because of this complication, mathematical models are needed to extract the average turnover rate of a population of cells from deuterium labeling data.

Deuterated Glucose Labeling

Because deuterium enrichment cannot be detected in lymphocytes during the first days of deuterated glucose labeling and labeling periods are generally very short, it is difficult to estimate the average turnover rate during the labeling phase in deuterated glucose labeling experiments. Instead, data collected during the post-labeling phase are typically used to estimate the average turnover rate. Because the loss of label post-labeling does not reflect the cell population as a whole, the average turnover rate is estimated via back-extrapolation of the post-labeling data to the moment of label cessation. Using this approach, information about the labeling phase can be obtained even though no data points were collected during this phase (Fig. 4a, b).

Heavy Water Labeling

In heavy water labeling experiments, the duration of heavy water administration permits frequent sampling both during the labeling and the post-labeling phase. The entire labeling curve (composed of both labeling and post-labeling data) can therefore yield reliable average production estimates (Fig. 4c, d).

Because body water enrichment is changing during the labeling period (it takes several days before body fluids reach a stable level of ^2H_2O enrichment) and after cessation of labeling (it takes several days before ^2H_2O is completely washed out), the fraction of labeled nucleotides in the dNTP pool progressively changes over time. Hence, cells that divide shortly after the start of labeling will not have the same probability of incorporation of labeled dNTPs as cells that divide when body water enrichment has reached a plateau. Conversely, during the post-labeling phase newly synthesized

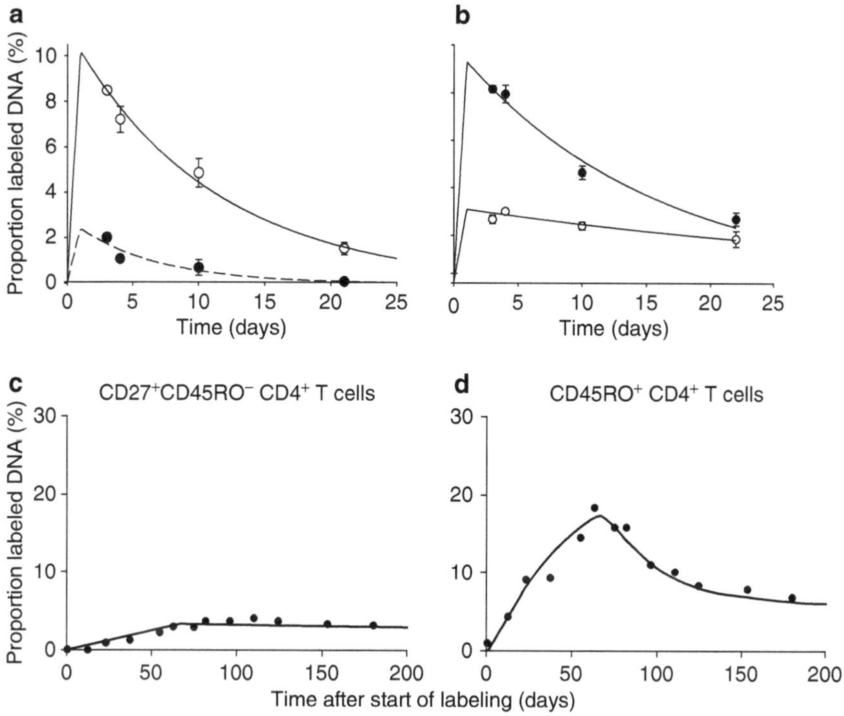

Fig. 4. **Typical lymphocyte subpopulation enrichment curves.** (**a**) and (**b**) Fraction of labeled DNA, taken as fraction of labeled cells (*F*, %) from DNA labeling of T-lymphocyte subpopulations for 1 day: (**a**) $CD4^{+}CD25^{bright}$ Regulatory T cells (open circles) with high turnover versus $CD4^{+}CD25^{-}$ cells (*filled circles*, *dashed line*), and (**b**) $CCR5^{+}$ (*filled circles*) and $CCR5^{-}$ (*open circles*) $CD45RO^{+}CD4^{+}$ T cells, from healthy young subjects following labeling with $^{2}H_2$-glucose. (Adapted from (15) and (37).) (**c**) Slowly dividing naive ($CD27^{+}CD45RO^{-}$) $CD4^{+}$ T cells, and (**d**) rapidly dividing memory ($CD45RO^{+}$) $CD4^{+}$ T cells from a healthy individual after labeling with heavy water for 9 weeks (*vertical bar*) (Adapted from (13)).

dNTPs will still become labeled as long as $^{2}H_2O$ has not been fully washed out. Therefore the loss of label does not simply reflect the disappearance rate of (labeled) cells, as dividing cells continue to incorporate label during the post-labeling phase. Mathematical modeling is essential to correct for the changing levels in body water enrichment during the entire sampling period (Fig. 1) (13).

2. Materials

For human studies, prior institutional and ethical approval should be obtained in accordance with local and national guidelines and regulations; informed consent must be obtained from all subjects before any interventions.

2.1. Labeling

2.1.1. Deuterium-Labeled Glucose

1. (6,6-$^{2}H_2$)-glucose (Cambridge Isotopes Inc or Isotec, Sigma-Aldrich), should be sterile and, if given intravenously (Option B), certified pyrogen-free (see Note 1).

2. Infusion fluids: water for injections or 0.45% saline for injection (Baxter Healthcare UK).

Option A—Oral administration

1. Weigh out 0.6–1.0 g/kg body weight deuterated glucose.
2. Make up to 240 ml final volume with water. Once dissolved use fresh or store at 4°C (see Note 2).

Option B—Intravenous administration

1. Prepare (6,6-$^2H_2$)-glucose infusate by taking 0.5–1 g/kg body weight (6,6-$^2H_2$)-glucose and reconstituting into 1,000 ml of 0.45% saline or water for injection (see Note 3). Withdraw approximately 200 ml, dissolve the glucose powder, and reinject into the infusate bag(s) through a 0.2 μm filter (because of volume expansion, the final volume will be increased). Prepare shortly before use and store at 4°C (see Note 3). Observe sterility precautions throughout to avoid contamination.

2.1.2. Deuterated Water

1. Deuterium oxide (D, 99.8%, Cambridge Isotope Laboratories). Follow local and national guidelines and regulations for testing: usual requirements would be for confirmation that the product meets requirements for purified water.
2. For the oral bolus, weigh out 10 ml deuterated water per kg body water (see Note 4). Body water is assumed to be 60% and 50% of body weight for males and females, respectively.
3. Maintenance dose = 1.25 ml deuterated water/kg body water per day. Subjects can take their deuterated water doses at home. Aliquot into amounts depending on the volume of their personal maintenance dose, the frequency of visits, and shelf-life.
4. Use fresh water; the shelf-life is determined locally and relates primarily to possible microbial contamination.

2.2. Sampling

2.2.1. Blood Withdrawal and Urine Sampling

1. Equipment for venepuncture and collection of samples: heparin-containing tubes, urine containers (for ^2H_2O labeling only), etc., and standard reagents for PBMC isolation by Ficoll density gradient centrifugation will be required.

2.2.2. Cell Separation Reagents

1. Reagents for magnetic or flow-cytometric separation will be required but are beyond the scope of this protocol.

2.3. Analysis

2.3.1. DNA Extraction Reagents

1. Use a proprietery DNA extraction kit, e.g., Qiagen QiaAmp Micro DNA extraction kit for small cell number samples; alternatively, use Macherey Nagel NUCLEOSPIN Blood Quick Pure or Qiagen DNeasy kit if cellular material/DNA is abundant, or Qiagen Flexigene kit for whole blood (baseline).

2.3.2. DNA Hydrolysis Reagents (Following Published Protocols (2))

1. Water, molecular biology grade.
2. 1 M NaOH, molecular biology grade.
3. Sodium acetate.
4. Acetic acid.
5. Zinc sulphate.
6. Acid phosphatase (potato, 1 kU; Calbiochem).
7. S1 nuclease (Sigma).

2.3.3. GC/MS Derivatization Reagents for DNA Analysis

Reagents marked * have significant toxicities—use protective equipment and follow local and national guidelines.

1. Pentafluorobenzyl hydroxylamine* (PFBHA, 1 mg/ml aqueous solution). Prepare solution fresh; store at 4°C for <1 week.
2. Acetic Anhydride* (Sigma-Aldrich).
3. N-methylimidazole* (Sigma-Aldrich). Store dry at 4°C.
4. Sodium sulphate, granular, anhydrous.
5. Dichloromethane* (Sigma-Aldrich).
6. Ethyl acetate (VWR International).

2.3.4. GC/MS Derivatization Reagents

Plasma glucose enrichment

1. Hydroxylamine hydrochloride* (Sigma-Aldrich).
2. Pyridine* (Sigma-Aldrich).

Body water enrichment

1. Calcium carbide (Sigma-Aldrich).

2.3.5. Standard Solutions

1. Unlabeled DNA (molecular biology grade, e.g., Calf thymus, Sigma-Aldrich).

Deuterated glucose labeling

1. DNA enrichment analysis: Combine (5,5-$^2H_2$)-2-deoxyribose (Cambridge Isotopes Inc.), which generates an (M+2) ion, with unlabeled deoxyribose (Sigma) to produce solutions of known proportions of labeled/unlabeled material covering the expected analytic range (0–1% atom percent express, APE=(M+2)/((M+0)+(M+2))).
2. Analysis of precursor (glucose) enrichment: Glucose (Sigma). Combine with (6,6-$^2H_2$)-glucose from Subheading 2.1.1 to prepare a standard curve of 0–50% enriched glucose.

Heavy water labeling

1. DNA enrichment analysis: Combine 1-^{13}C-deoxyadenosine (Cambridge Isotopes Inc.), which generates an (M+1) ion, with unlabeled deoxyadenosine (Sigma) in known proportions

to cover the expected analytic range (0–10% APE for water-labeling).

2. Analysis of precursor (body water) enrichment.

Combine ^2H_2O (see Subheading 2.1.2) with H_2O to prepare standards with enrichment 0.25%, 0.5%, 1%, 2% (volume to volume ratio), and higher if the experimental range is larger. As controls, use urine to which ^2H_2O is added in multiple concentrations.

2.3.6. Equipment

1. Heat block to dry under nitrogen gas, "Speedvac" or equivalent sample concentrator.
2. Cell sorting equipment.
3. Gas Chromatograph Mass Spectrometer (GC/MS; Agilent 5973/6890 with HP-225 or DB-17 column, Agilent Technologies, or equivalent).

Deuterated glucose labeling

1. Lancets.
2. Filter paper.

Option B—Intravenous administration

1. 0.2 μm filters (Sartorius Stedim Biotech GmbH).
2. Infusion pump (IVAC 590 volumetric pump, or equivalent), calibrated gravimetrically (see Note 5).
3. Canula and intravenous administration sets (Venflon, BD Medical or equivalent).

Heavy water labeling

1. GC/MS equipment with column: Porabond Q, 25 m × 0.32 mm × 15 μm (Chrompack).

3. Methods

3.1. Labeling

Remember to take baseline samples (Subheading 3.2.1) before labeling.

3.1.1. Deuterated Glucose

Option A—Oral administration

1. Take oral glucose solution (Subheading 2.1—option A); concentration should be about 200 g/l and volume 240 ml. Administer 36 ml oral glucose solution at time zero (T0). Aliquot the glucose solution into a disposable cup from which the subject drinks. Follow with >2 rinses of the cup with water, each of which the subject should drink. Repeat for all doses.

2. Administer further 10 ml doses every half-hour thereafter until 10 h later (T10). Meals should be restricted to ≤200 kcal (low glycaemic index foods are preferred) to avoid large changes in glucose enrichment. Commercially available "diet" meals that typically comprise about 200 kcal each (~15 g carbohydrate, ~8 g fat) are suitable. Meals may be given every 2–3 h. Discourage physical activity.
3. Monitor glucose enrichment with finger-prick blood samples, as below, Subheading 3.2.2.

Option B—Intravenous administration

1. Insert intravenous canula—can also be used for baseline blood sample (Subheading 3.2.1); flush canula with 0.9% saline.
2. Take intravenous glucose solution (Subheading 2.1—Option B); concentration should be about 60 g/l and volume about 1,100 ml (see Note 2).
3. Set up infusion equipment. Start infusion at 300 ml/h and record the exact start time.
4. Run at 300 ml/h for 15 min (75 ml), as priming dose, then at ~43 ml/h thereafter (see Note 6).
5. Administer small regularly-spaced meals. We typically give four meals (each ≤200 kcal) at ~3, 6, 9, and 12 h with a snack at 15 h and a smaller meal at 23 h. Discourage excessive physical activity.
6. Monitor subject body temperature, pulse and blood pressure at least every 4 h (see Note 7).
7. Monitor glucose enrichment with finger-prick blood samples (see Subheading 3.2.2).
8. At the end of infusion, record the exact time. This is needed to calculate the area under curve of glucose enrichment versus time, against which DNA labeling is compared (Subheading 3.3.1, step 17).

3.1.2. Heavy Water

Remember to take baseline samples (Subheading 3.2.1) before labeling.

1. Calculate the volume of the start bolus (typically between 200 and 400 ml) based on the weight of the participant (as in Subheading 2.1.2). Divide into aliquots, for half-hourly administration, distributed evenly over the day. Phased intake is important to reduce the occurrence/severity of dizziness due to transient side effects on the vestibular apparatus (see Note 8). Every half hour, aliquot one portion into a disposable cup from which the subject drinks. Do this until the last dose is

taken (see Note 8). Subjects are encouraged to have breakfast and eat and drink normally during the day.

2. Monitor the increase in body water enrichment resulting from the start bolus by taking a urine sample after administration of the last portion of the start bolus (see Subheading 3.2.2, body water enrichment).
3. After administration of the bolus, subjects are given heavy water to drink their daily maintenance dose at home for the indicated period of time.

3.2. Sampling

3.2.1. Baseline Samples

1. Take baseline blood sample by venepuncture (*before labeling*); aliquot for tests as follows:
   (a) Baseline DNA enrichment—heparinized whole blood or sorted cells (the baseline DNA sample may be processed straightaway or frozen at −20°C and processed later as described in Subheading 3.3.2).
   (b) Complete blood count/differential lymphocyte count to aid interpretation of results.
   (c) Other clinical samples, e.g., biochemistry, serology (~5 ml) as required for clinical interpretation.
   (d) For glucose labeling: Baseline glucose enrichment—3–4 blood spots on filter paper; air-dry then store at 4°C until analysis.

3.2.2. Monitoring Precursor Enrichment

Glucose enrichment

1. Pinprick carefully cleaned finger/thumb and blot 3–4 spots onto filter paper at time-points: 1, 4, 7, and 10 h for Option A (oral administration), or 1, 4, 8, 12, 20 and 23 h for Option B (intravenous administration), in addition to baseline (T0), already taken (see Subheading 3.2.1) (see Note 9).
2. Leave to air-dry (≥10 min).
3. Store dried blood spots on filter paper at 4°C until glucose extraction.

Body water enrichment

1. Take the first urine sample directly after the last aliquot of the oral bolus (see Subheading 3.1.2). After this, it is recommended and sufficient to sample once or twice a week from start of labeling until a steady-state level is reached (usually within the first 3 weeks of labeling), and from the end of labeling until heavy water is completely washed out (usually within the first 3 weeks post-labeling). Depending on the length of the plateau phase during labeling, urine can be collected during this phase once a week or less, to ensure that steady state levels are maintained. For example, in a 9 week labeling experiment, urine can be collected once or

twice in week 1 and 2 and once in week 3, 4, 5, 6, 7, and 9 during labeling, and post-labeling once or twice in week 10 and 11, and once in week 12, 13 and 14 of the study. Urine samples can easily be collected at home, reducing the number of visits and can be frozen at −20°C until further analysis (see Subheading 3.3.1).

3.2.3. Blood Sampling and Cell Sorting to Monitor DNA Deuterium Enrichment

1. Take follow-up blood samples (~50 ml; volume will depend upon the frequency of the cell of interest and analytic sensitivity) into preservative-free heparin (≥20 U/ml blood) collection tubes (alternative tubes may be suitable for some cell-sorting procedures).
   (a) For glucose labeling, e.g., at 3, 4, 10, and 21 days post-labeling.
   (b) For ^2H_2O labeling, at frequent intervals over the following weeks, both during and after labeling. For example, after 1, 2, 3, 4, 6, and 9 weeks (during labeling) and after 10, 11, 12, 14, 16, 21, and 30 weeks of the start of the study (post-labeling).
2. Isolate PBMC by density centrifugation on Ficoll-Paque (Amersham Biosciences) and sort cell populations of interest (see Note 10).
3. Proceed to DNA extraction (Subheading 3.3.2) or store separated cells frozen at −70°C until DNA extraction (see Note 10).

3.3. Analysis

3.3.1. Precursor Enrichment Analysis

Blood glucose enrichment

1. This is achieved by first extracting glucose from filter paper blood spots, then derivatizing to the aldonitrile triacetate (ATA) derivative and finally GC/MS analysis (see Note 11).
2. Cut at least two blood drops on filter paper from Subheading 3.2.2 into a 1.5 ml centrifuge tube and add 1 ml 50% ethanol. To avoid contamination, wear gloves and use designated scissors, cleaning blades between samples (see Note 9).
3. Leave at ambient temperature (approximately 20°C) for at least 30 min.
4. Vortex and transfer supernatant to another microcentrifuge tube.
5. Centrifuge (>15,000×*g*, 10 min, ambient temperature, in a 1.5 ml centrifuge tube) to remove any precipitate; transfer to a clean tube and dry under nitrogen gas at 50°C. Ethanol extract can be stored at −20°C until derivatization.
6. Make a fresh solution of 1% w/v hydroxylamine·HCl in pyridine (1 mg/100 μl).
7. To the dried samples from step 5, and to standards of glucose of known enrichment (from Subheading 2.3.5), add 25 μl of hydroxylamine/pyridine reagent, seal and mix gently.

8. Heat the samples at 100°C for 60 min in a dry heat block.
9. Cool and quick spin the samples in a microfuge.
10. At ambient temperature, add 25 μl of acetic anhydride and seal the tube. Mix gently.
11. Incubate at room temperature for 30 min. To ensure completion of the derivatization reaction, heat for the last 10 min at 70°C in a heat block.
12. Quick spin the samples in a microfuge, to ensure all the solution is at the bottom of the tube, then dry at 50°C under nitrogen.
13. Resuspend the derivatized samples and standards in 400 μl of ethylacetate.
14. Vortex briefly, then pulse-microfuge to remove any particulate matter or precipitate, transferring the supernatant to vials ready for analysis by GC/MS.
15. Analyze by GC/MS, monitoring in SIM mode for ions *m/z* 328 (M+0) and 330 (M+2). We use an Agilent 5973/6890 with HP-225 column (Agilent Technologies) (see Note 12).
16. Determine enrichment from the ratio of ions (M+2)/((M+0)+(M+2)), calibrating against standard glucose samples of known enrichment from Subheading 2.3.5.
17. Calculate the area under curve (AUC) for glucose enrichment versus time by the trapezoid method (5, 7) (see Note 13). The corrected value gives the mean glucose × time enrichment value.

Body water enrichment

This section of the protocol follows the method described by Van Kreel et al. (33) by converting calcium carbide (CaC_2) and urine- or plasma-derived water to acetylene (C_2H_2), with minor modifications. During this reaction, deuterium atoms from water are transferred to acetylene. Measured enrichment in acetylene is equal to the enrichment in urine or plasma (hence body water).

1. Weigh 30 mg powdered calcium carbide (CaC_2) in each GC vial.
2. Cap each vial and evacuate using a needle connected to vacuum.
3. Pipette 2 μl of urine (or plasma) through the cap on the calcium carbide using a microliter syringe (see Note 14).
4. Include a blank (distilled water), standards and controls.
5. Incubate for 5 min.
6. The gas inside the vials can now be injected into the GC/MS.
7. Analyze by GC/MS using selective ion monitoring (SIM) quantifying ions *m/z* 26 (M+0) and *m/z* 27 (M+1) for H_2O and ^2H_2O, respectively.

3.3.2. DNA Enrichment Analysis

This section of the protocol follows the method described by Busch et al. (2) digesting DNA to release nucleosides, then derivatizing to the pentafluoro tri-acetate (PFTA) derivative before GC/MS.

1. Take sorted cells from Subheading 3.2.3 (including baseline) and extract DNA using a commercial kit (see Note 15). If using the Qiagen QiaAmp Micro DNA extraction kit, follow the protocol modifications described by Busch et al. (2); specifically, at the end of extraction, DNA should be suspended in 200 μl water, not buffer. Store DNA at −20°C or proceed directly to hydrolysis, (see next step). The minimum number of cells will depend upon local analytic sensitivity and should be derived prior to defining blood volumes required for the subpopulation of interest. Alternatively, sorted cell pellets can be boiled at 100°C in 200 uls of water for one hour and put on ice immediately for 10 minutes.
2. To digest the extracted DNA samples, or boiled samples, as described by Busch et al. (2), add 50 μl of hydrolysis cocktail and incubate at 37°C overnight with shaking.
3. Derivatize the digested DNA samples to nucleoside pentafluoro tri-acetate (PFTA) derivatives as follows, alongside (5,5-$^{2}H_{2}$)-ribose (for M+2) or 1-^{13}C-deoxyadenosine (for M+1) standard solutions of known enrichment from Subheading 2.3.5 to calibrate isotope ratios.
4. Transfer the digested samples into 16 mm×100 mm screw-capped glass tubes. To each sample/standard add:
   (a) 100 μl of freshly made aqueous pentafluorobenzyl hydroxylamine solution (1 mg/ml).
   (b) 75 μl of glacial acetic acid.
5. Cap the tubes and incubate on heating block for 30 min at 100°C.
6. Remove samples and allow to cool to room temperature.
7. Add 1 ml of acetic anhydride, then 100 μl of N-methylimidazole, mixing immediately. Perform this step in a fume hood, wearing protective goggles and pointing the opening of the tube away from you. Samples may splash as the exothermic acetylation reaction proceeds due to sudden overheating. Allow the reaction to proceed at ambient temperature for 15–20 min during which the samples will cool down.
8. Add 2 ml of water to the reactions, vortex for 10 s.
9. Add 750 μl of dichloromethane to the tubes and vortex for 5 s. Allow phases to separate (~1 min).

10. Transfer 500 μl of the bottom (organic) layer (from step 9) into 2 ml microcentrifuge tubes. Avoid transferring any of the aqueous phase, which may introduce contaminants. Wet the pipette tip with dichloromethane before transfer to help reduce inadvertent mixing.
11. Add a further 750 μl of dichloromethane to the reaction tubes and repeat the dichloromethane extraction (vortex for 5 s, then allow phases to separate, ~1 min), adding the organic layer to that already extracted.
12. Dry the microcentrifuge tubes — we use content using a SpeedVac at ambient temperature for at least 4 h, or ideally overnight. Leave the heat setting "off" as heating may cause evaporative loss of some derivative. Verify complete drying visually; avoid residual moisture or acid as this may damage the GC column. Drying in a stream of nitrogen is not sufficient to remove residual acetic acid.
13. Resuspend each sample in 250 μl of ethyl acetate, vortex and pulse centrifuge to remove any precipitate or solid material. Transfer the supernatant to a GC glass insert. Avoid transfering any precipitate or solid material.
14. Evaporate ethyl acetate (SpeedVac, ambient temperature, ~1 h).
15. Resuspend each sample in a volume of ethyl acetate suitable for the GC/MS autosampler (typically 50 μl). Place the glass insert into a labeled GC vial and cap immediately.
16. Analyze by GC/MS using selective ion monitoring (SIM) quantifying ions *m/z* 435 (M+0), and *m/z* 437 (M+2) for deuterated glucose or 436 (M+1) for heavy water labeling (see Note 16). Use the isotope ratio of abundance-matched samples to calculate the enrichment of deuterated deoxyadenosine, calibrating against standard curves of known enrichment from Subheading 2.3.5 (see Note 17).

3.4. Interpretation

Data analysis: deduction of cellular production and disappearance rates

Different models have been proposed to analyze enrichment data from stable isotope labeling experiments. The most appropriate model depends on the exact biology of the cells under investigation (see Note 18). Here we present a basic model that is applicable to many circumstances, and which captures the general principles of a stable isotope labeling experiment. Models like these need to be fitted to the labeling data to obtain biological parameters such as the average turnover rate of a lymphocyte population.

1. Express the enrichment data as the ratio of labeled to total PFTA-nucleoside derivative (predominantly arising from

deoxyadenosine, although all purine nucleosides should be similarly labeled). This gives the fraction of labeled DNA fragments $L(t)$. Plot the enrichment data as a function of time t. Enrichment should start at zero at baseline, rise to a peak (at or after label cessation), then fall thereafter.

2. Independent of the source of stable isotope that is used (i.e., deuterated glucose or heavy water), the change in the fraction of labeled DNA fragments in a population of cells over time can be described by the following differential equation:

$$\mathrm{d}L(t) / \mathrm{d}t = pc\ P(t) - d^* L \qquad (1)$$

where p is the average production rate of the cells, d^* the loss rate of labeled cells, and $c\ P(t)$ is the chance that an incorporated deoxyadenosine is labeled (see Note 18). The latter term is a composite between the availability of deuterium (the precursor) $P(t)$ which changes over time, and an amplification factor c accounting for the fact that there are multiple hydrogen atoms in a single DNA fragment that can be replaced by deuterium (13).

3. The curve described by Eq. 1. is fitted to the enrichment data using nonlinear least squares regression. This can be done with a number of software packages that allow user-defined equations, such as *R* or *Sigmaplot* (Systat Software Inc). Typical data plots and curve-fitting are shown in Fig. 4.

4. The half-life, $t^{1/2}$, of the lymphocytes under investigation can be calculated as $\ln 2/p$. Note that p will include proliferation at all sites, e.g., for naive T cells it will include not only peripheral T-cell proliferation but also production of new lymphocytes by the thymus.

Deuterated glucose labeling

1. Because of the fast turnover of glucose, during the labeling phase, $P(t)$ can be assumed to be continuously at its maximal level, while after label cessation $P(t)=0$. Precursor enrichment during labeling is typically measured as the label enrichment in plasma glucose multiplied by a factor 0.73 to account for intracellular label dilution (4) (see Note 13).

2. In most deuterated glucose labeling experiments, label is given so briefly that enrichment data during the labeling phase are lacking. In that case, the average turnover rate of the cell population, p, is estimated by fitting the loss of label after label cessation to the solution of Eq. 1 with $P(t)=0$ (11, 12). The loss curve is subsequently back-extrapolated to the moment of label cessation in order to estimate p. If deuterated glucose is

given for longer periods, proceed as under *Heavy water labeling*, step 2.

Heavy water labeling

1. Because the turnover rate of water is relatively slow, changes in the availability of deuterium need to be explicitly taken into account. This can be done by fitting a simple exponential accrual and loss function to the enrichment data from urine:

$$P(t) = f(1 - e^{-\varepsilon t}) + \beta e^{-\varepsilon t} \text{during label administration} \quad (2a)$$

and

$$P(t) = (f(1 - e^{-\varepsilon T}) + \beta e^{-\varepsilon T})e^{-(\varepsilon t - T)} \text{after label cessation} \quad (2b)$$

where f represents the fraction of deuterated water in the drinking water, ε the turnover rate of body water per day, and β the body water enrichment that is attained after the boost of label by the end of day 0.

2. The best fit of the function $P(t)$ is substituted into Eq. 1, after which the latter function is fitted to (a) the level of label incorporation in a population of cells that turn over very rapidly, such as granulocytes, so as to estimate the value of c, and subsequently to (b) the enrichment data of the cell population of interest during and after the labeling phase, so as to estimate its turnover parameters p and d^*.

4. Notes

Parentheses demonstrate whether the note refers to glucose or heavy water labeling.

1. (gluc) (6,6-$^2H_2$)-glucose should be tested for sterility and pyrogenicity and certified by the manufacturer. Microbial contamination of the infusate must be avoided. Reconstitution as an infusate must be performed under sterile conditions. Passage of reconstituted glucose through a 0.2 μm filter is advisable to ensure sterility.
2. (gluc) Once reconstituted, use fresh or store at 4°C. Glucose is chemically stable; the shelf-life is determined by local pharmacy guidelines and relates primarily to possible microbial contamination.
3. (gluc) Hyperosmolar solutions may cause phlebitis. The final osmolarity of the solution will be about 450 mOsmol/L; this should not cause phlebitis. Avoid using infusion volumes of less than 1,000 ml.

4. (water) Because the administration of the bolus of deuterated water during the first day can lead to vertigo and/or nausea, one can lower the dose to, for example, 7.5 ml deuterated water per kg body water to minimize side effects. A draw-back is that the maximal level of label enrichment in the body water is attained later, but more frequent sampling of urine during the labeling phase will help to obtain the necessary information of how much label was available at any time-point. See also Note 8.
5. (gluc) Infusion pumps may not always run at the stated rate and should be calibrated gravimetrically prior to research use.
6. (gluc) A priming dose of about 1.8 times the hourly dose was found empirically to reach plateau levels in most subjects; under-priming or over-priming results in rising or falling glucose enrichments in plasma (respectively) and should be avoided. Glucose enrichment tends to rise overnight as feeding is not continued during this time. Such curves can be converted to a "square wave" by calculating AUC (see Note 13) and expressing this as the mean enrichment achieved throughout the duration of label administration. However, results may be slightly biased either towards the earlier or later part of the day; this is not considered a significant bias.

   The infusion rate may need to be modified empirically as it will depend upon the infusion pump characteristics (which should be calibrated) and because the volume of the infusate is difficult to predict; bags often contain slightly more than the nominal volume to allow for line priming, and because adding glucose expands the volume, depending upon the amount added. Aim to give the whole dose over 24 h.
7. (gluc) In order to monitor for the possibility of a pyrogen or infusion reaction, subjects should be closely monitored by a clinically trained staff member and pulse, temperature, and blood pressure recorded at least every 4 h.
8. (water) Transient nausea or vertigo during intake of the initial bolus of ^2H_2O has been attributed to the density of heavy water being higher than the density of the endolymph fluid, which creates a transient density gradient to which the hair cells of the inner ear are sensitive (27). Phased administration of the bolus in small portions spread over the day is usually effective in preventing or strongly reducing side effects on the inner ear. However, if individuals experience more severe and/or longer lasting side effects, it is possible to postpone the next dose of ^2H_2O and have longer intervals between doses, or divide the bolus into even smaller doses. If side effects persist, the ramp-up protocol can be discontinued. In this case, body water enrichment will plateau later (see also Note 4). Intake of the

daily maintenance dose has never been associated with any side effects.

9. (gluc) Aberrant or unexpectedly high glucose enrichment levels in blood spots may indicate contamination. Always plot the glucose time-profile. Avoid any possible contamination from the infusate or oral solution onto filter paper blood spots. This is a particular risk with the oral solution, which is easy to transfer from mouth or cup via hand to filter paper and is highly concentrated such that even trivial contamination will give aberrant results. Use disposable cups; take care that both the subject and operator wear gloves when handling oral glucose solution even though it is nonhazardous. Clean the subject's finger carefully prior to pinprick blood testing.
10. (both) We recommend sorting cells fresh. It may be possible to freeze the cells at –70°C prior to sorting. If cells are frozen, it needs to be established that freezing does not preferentially affect the recovery of the labeled cells of interest. We also recommend commencing DNA extraction from fresh cells by adding lysis buffer straight after sorting rather than storing as cell pellets. Note also that the use of intracellular staining protocols for cell sorting may reduce DNA yield.
11. (gluc) There are many published alternative derivatization and GC/MS protocols for analysis of isotopic enrichment of glucose.
12. (both) Detailed GC/MS protocols are beyond the scope of this article, but note that optimization of chromatography is crucial for reliable quantitation.
13. (gluc) Calculation of the area under curve (AUC) for glucose enrichment versus time by the trapezoid method applies the equation:

$$\mathrm{AUC} = [E_0 + E_1] \times t_1 / 2 + [E_1 + E_2] \times (t_2 - t_1) / 2 + [E_2 + E_3] \times (t_3 - t_2) / 2 + \ldots t_n + \mathrm{AUC}_{\mathrm{add}}$$

where E is the glucose enrichment at times, 0, 1, 2, 3 …n, and $\mathrm{AUC}_{\mathrm{add}}$ is an estimated additional post-infusion area ($\mathrm{AUC}_{\mathrm{add}}$) in order to allow for additional labeling that occurs during the die-away of labeled glucose. For a full discussion see Macallan et al. (5) and Vukmanovic-Stejic et al. (7).

Dividing this sum by the infusion/administration time gives the mean glucose enrichment, E_{Glu}. Multiply this by the correction factor derived for this cell type to obtain the mean precursor enrichment. (We have used 0.65 previously, (5) but more recent studies suggest a value of 0.73 is appropriate for dividing lymphocytes; Kovacs et al. have used a slightly lower value of 0.60 in their studies (24, 25, 34).

14. (water) Carbide and water form an explosive mixture. Do not weigh too much carbide in to the GC vials and do not pipette more than 2 μl of water onto the carbide.
15. (both) Alternative DNA extraction protocols and kits are available and may be used.
16. (both) Detailed gas chromatography and mass spectrometry (GC/MS) analysis protocols for DNA from deuterium labeling studies have been described in Busch et al. (2). An alternative microcapillary liquid chromatography-electrospray ionization (LC-ESI)/MS has been described (30), an advantage of which is that it does not require a derivatization step. However, it may require larger sample amounts for quantitation.
17. (both) Isotope ratios are susceptible to abundance artifacts. Run samples as single injections first, to check abundance; then dilute or adjust injection volume to ensure equal abundances are achieved for all samples and standards; then repeat analysis of ratios in duplicate or triplicate.
18. (both) Most models assume that the size of the cell population under investigation is constant during the labeling protocol. Nevertheless, the average turnover rate p of the cell population does not necessarily equal the loss rate of labeled cells d^*. Equations can be adapted to model cell population sizes that are not constant over time.

    The production rate p includes both production of the cells of interest within the specific cell population as well as any immediate precursors which divided in the presence of label and subsequently matured or trafficked to the pool of interest. The disappearance rate d^* includes cell death, net trafficking of cells out of the peripheral blood and disappearance due to phenotype switching.

    This model has been widely applied (13, 17, 35), but it should be noted that this is not the only model available. For a review of alternative models see Borghans et al. (14). For example, specific models have been developed to describe populations of cells that are kinetically heterogeneous, i.e., that consist of sub populations of cells with different rates of turnover (36).

Acknowledgements

We acknowledge financial support from the Medical Research Council (UK), the Wellcome Trust, the Charitable Trustees of St George's Hospital, London, and the Netherlands Organization for Scientific Research (NWO, grants 917.96.350 and 836.07.002) during the execution of studies included in this report.

References

1. Asquith B, Borghans JA, Ganusov VV, Macallan DC (2009) Lymphocyte kinetics in health and disease. Trends Immunol 30:182–189
2. Busch R, Neese RA, Awada M, Hayes GM, Hellerstein MK (2007) Measurement of cell proliferation by heavy water labeling. Nat Protoc 2:3045–3057
3. Hellerstein MK (1999) Measurement of T-cell kinetics: recent methodologic advances. Immunol Today 20:438–441
4. Macallan DC, Fullerton CA, Neese RA, Haddock K, Park S, Hellerstein MK (1998) Measurement of cell proliferation by labeling of DNA with stable isotope-labeled glucose: studies in vitro, in animals and in humans. Proc Natl Acad Sci U S A 95:708–713
5. Macallan DC, Asquith B, Zhang Y, de Lara CM, Ghattas H, Defoiche J, Beverley PC (2009) Measurement of proliferation and disappearance of rapid turnover cell populations in human studies using deuterium-labelled glucose. Nat Protoc 4:1313–1327
6. Neese RA, Misell LM, Turner S, Chu A, Kim J, Cesar D, Hoh R, Antelo F, Strawford A, McCune JM, Christiansen M, Hellerstein MK (2002) Measurement in vivo of proliferation rates of slow turnover cells by ^2H_2O labeling of the deoxyribose moiety of DNA. Proc Natl Acad Sci U S A 99:15345–15350
7. Vukmanovic-Stejic M, Zhang Y, Akbar AN, Macallan DC (2011) Measurement of proliferation and disappearance of regulatory T cells in human studies using deuterium-labeled glucose. Methods Mol Biol 707:243–261
8. Hellerstein M (2000) Methods for measuring polymerisation biosynthesis: three general solutions to the problem of the "true precursor". Diabetes Nutr Metab 13:46–60
9. Cohen A, Barankiewicz J, Lederman HM, Gelfand EW (1983) Purine and pyrimidine metabolism in human T lymphocytes. Regulation of deoxyribonucleotide metabolism. J Biol Chem 258:12334–12340
10. Reichard P (1988) Interactions between deoxyribonucleotide and DNA synthesis. Annu Rev Biochem 57:349–374
11. Asquith B, Debacq C, Macallan DC, Willems L, Bangham C (2002) Lymphocyte kinetics: the interpretation of labelling data. Trends Immunol 23:596–601
12. Macallan DC, Asquith B, Irvine A, Wallace D, Worth A, Ghattas H, Zhang Y, Griffin GE, Tough D, Beverley PC (2003) Measurement and modeling of human T cell kinetics. Eur J Immunol 33:2316–2326
13. Vrisekoop N, den Braber I, de Boer AB, Ruiter AF, Ackermans MT, van der Crabben SN, Schrijver EH, Spierenburg G, Sauerwein HP, Hazenberg MD, De Boer RJ, Miedema F, Borghans JA, Tesselaar K (2008) Sparse production but preferential incorporation of recently produced naive T cells in the human peripheral pool. Proc Natl Acad Sci U S A 105:6115–6120
14. Borghans JA, De Boer RJ (2007) Quantification of T-cell dynamics: from telomeres to DNA labeling. Immunol Rev 216:35–47
15. Vukmanovic-Stejic M, Zhang Y, Cook JE, Fletcher JM, McQuaid A, Masters JE, Rustin MH, Taams LS, Beverley PC, Macallan DC, Akbar AN (2006) Human CD4+ CD25hi Foxp3+ regulatory T cells are derived by rapid turnover of memory populations in vivo. J Clin Invest 116:2423–2433
16. Macallan DC, Wallace DL, Irvine A, Asquith B, Worth A, Ghattas H, Zhang Y, Griffin GE, Tough DF, Beverley PC (2003) Rapid turnover of T cells in acute infectious mononucleosis. Eur J Immunol 33:2655–2665
17. Macallan DC, Wallace D, Zhang Y, De Lara C, Worth AT, Ghattas H, Griffin GE, Beverley PC, Tough DF (2004) Rapid turnover of effector-memory CD4(+) T cells in healthy humans. J Exp Med 200:255–260
18. Wallace DL, Zhang Y, Ghattas H, Worth A, Irvine A, Bennett AR, Griffin GE, Beverley PC, Tough DF, Macallan DC (2004) Direct measurement of T cell subset kinetics in vivo in elderly men and women. J Immunol 173:1787–1794
19. Macallan DC, Wallace DL, Zhang Y, Ghattas H, Asquith B, De Lara C, Worth A, Panayiotakopoulos G, Griffin GE, Tough DF, Beverley PC (2005) B cell kinetics in humans: rapid turnover of peripheral blood memory cells. Blood 105:3633–3640
20. Defoiche J, Debacq C, Asquith B, Zhang Y, Burny A, Bron D, Lagneaux L, Macallan DC, Willems L (2008) Reduction of B cell turnover in chronic lymphocytic leukaemia. Br J Haematol 143:240–247
21. Hellerstein M, Hanley MB, Cesar D, Siler S, Papageorgopoulos C, Wieder E, Schmidt D, Hoh R, Neese R, Macallan D, Deeks S, McCune JM (1999) Directly measured kinetics of circulating T lymphocytes in normal and HIV-1-infected humans. Nat Med 5:83–89
22. Hellerstein MK, Hoh RA, Hanley MB, Cesar D, Lee D, Neese RA, McCune JM (2003) Subpopulations of long-lived and short-lived T

cells in advanced HIV-1 infection. J Clin Invest 112:956–966

23. McCune JM, Hanley MB, Cesar D, Halvorsen R, Hoh R, Schmidt D, Wieder E, Deeks S, Siler S, Neese R, Hellerstein M (2000) Factors influencing T-cell turnover in HIV-1-seropositive patients. J Clin Invest 105:R1–R8
24. Kovacs JA, Lempicki RA, Sidorov IA, Adelsberger JW, Sereti I, Sachau W, Kelly G, Metcalf JA, Davey RT Jr, Falloon J, Polis MA, Tavel J, Stevens R, Lambert L, Hosack DA, Bosche M, Issaq HJ, Fox SD, Leitman S, Baseler MW, Masur H, Di MM, Dimitrov DS, Lane HC (2005) Induction of prolonged survival of CD4+ T lymphocytes by intermittent IL-2 therapy in HIV-infected patients. J Clin Invest 115:2139–2148
25. Read SW, Lempicki RA, Di MM, Srinivasula S, Burke R, Sachau W, Bosche M, Adelsberger JW, Sereti I, Davey RT Jr, Tavel JA, Huang CY, Issaq HJ, Fox SD, Lane HC, Kovacs JA (2008) CD4 T cell survival after intermittent interleukin-2 therapy is predictive of an increase in the CD4 T cell count of HIV-infected patients. J Infect Dis 198:843–850
26. Mohri H, Perelson AS, Tung K, Ribeiro RM, Ramratnam B, Markowitz M, Kost R, Hurley A, Weinberger L, Cesar D, Hellerstein MK, Ho DD (2001) Increased turnover of T lymphocytes in HIV-1 infection and its reduction by antiretroviral therapy. J Exp Med 194:1277–1287
27. Brandt T (1990) Positional and positioning vertigo and nystagmus. J Neurol Sci 95:3–28
28. Van Gent R, Kater AP, Otto SA, Jaspers A, Borghans JA, Vrisekoop N, Ackermans MA, Ruiter AF, Wittebol S, Eldering E, van Oers MH, Tesselaar K, Kersten MJ, Miedema F (2008) In vivo dynamics of stable chronic lymphocytic leukemia inversely correlate with somatic hypermutation levels and suggest no major leukemic turnover in bone marrow. Cancer Res 68:10137–10144
29. Messmer BT, Messmer D, Allen SL, Kolitz JE, Kudalkar P, Cesar D, Murphy EJ, Koduru P, Ferrarini M, Zupo S, Cutrona G, Damle RN, Wasil T, Rai KR, Hellerstein MK, Chiorazzi N (2005) In vivo measurements document the dynamic cellular kinetics of chronic lymphocytic leukemia B cells. J Clin Invest 115:755–764
30. Fox SD, Lempicki RA, Hosack DA, Baseler MW, Kovacs JA, Lane HC, Veenstra TD, Issaq HJ (2003) A comparison of microLC/electrospray ionization-MS and GC/MS for the measurement of stable isotope enrichment from a ($^2H_2$)-glucose metabolic probe in T-cell genomic DNA. Anal Chem 75:6517–6522
31. Voogt JN, Awada M, Murphy EJ, Hayes GM, Busch R, Hellerstein MK (2007) Measurement of very low rates of cell proliferation by heavy water labeling of DNA and gas chromatography/pyrolysis/isotope ratio-mass spectrometric analysis. Nat Protoc 2:3058–3062
32. Neese RA, Siler SQ, Cesar D, Antelo F, Lee D, Misell L, Patel K, Tehrani S, Shah P, Hellerstein MK (2001) Advances in the stable isotope-mass spectrometric measurement of DNA synthesis and cell proliferation. Anal Biochem 298:189–195
33. Van Kreel BK, van der Vegt F, Meers M, Wagenmakers T, Westerterp K, Coward A (1996) Determination of total body water by a simple and rapid mass spectrometric method. J Mass Spectrom 31:108–111
34. Stevens RA, Lempicki RA, Natarajan V, Higgins J, Adelsberger JW, Metcalf JA (2006) General immunologic evaluation of patients with human immunodeficiency virus infection. In: Detrick B, Hamilton RG, Folds JD (eds) Manual of molecular and clinical laboratory immunology. ASM Press, Washington, DC, pp 848–861
35. Asquith B, Zhang Y, Mosley AJ, de Lara CM, Wallace DL, Worth A, Kaftantzi L, Meekings K, Griffin GE, Tanaka Y, Tough DF, Beverley PC, Taylor GP, Macallan DC, Bangham CR (2007) In vivo T lymphocyte dynamics in humans and the impact of human T-lymphotropic virus 1 infection. Proc Natl Acad Sci U S A 104:8035–8040
36. Ganusov VV, Borghans JA, De Boer RJ (2010) Explicit kinetic heterogeneity: mathematical models for interpretation of deuterium labeling of heterogeneous cell populations. PLoS Comput Biol 6:e1000666
37. Macallan DC, Zhang Y, De Lara C, Worth A, Beverley PC (2011) CD4+ T cell turnover is related to chemokine receptor expression and HIV viral co-receptor tropism. Proceedings of 17th conference on retroviruses and opportunistic infections abstract #270.

Chapter 11

Real-Time Quantitative (RQ-)PCR Approach to Quantify the Contribution of Proliferation to B Lymphocyte Homeostasis

Menno C. van Zelm, Magdalena A. Berkowska, Mirjam van der Burg, and Jacques J.M. van Dongen

Abstract

The cells of the adaptive immune system, B and T lymphocytes, each generate a unique antigen receptor through V(D)J recombination of their immunoglobulin (Ig) and T-cell receptor (TCR) loci, respectively. Such rearrangements join coding elements to form a coding joint and delete the intervening DNA as circular excision products containing the signal joint. These excision circles are relatively stable structures that cannot replicate and have no function in the cell. Since the coding joint in the genome is replicated with each cell division, the ratio between coding joints and signal joints in a population of B cells can be used as a measure for proliferation. This chapter describes a real-time quantitative polymerase chain reaction (RQ-PCR)-based approach to quantify proliferation through calculating the ratio between coding joints and signal joints of the frequently occurring kappa-deleting rearrangements in the *IGK* light chain loci in man and mouse. The approach is useful to study the contribution of proliferation to B-cell homeostasis in health and disease.

Key words: B cell, Immunoglobulin, V(D)J recombination, KREC, Replication history, Proliferation, Real-time quantitative PCR, IntronRSS, Kde, *IGK*

1. Introduction

B cells play a unique role in the vertebrate immune system, being responsible for the production of antibodies (immunoglobulin; Ig) that are capable of neutralizing invading pathogens. Importantly, antibody responses are highly antigen-specific and generate immunological memory: upon secondary encounter, the same pathogen is eliminated much more efficiently. These adaptive capabilities are the result of two independent stages of development (1). First, each B cell creates a unique antibody during antigen-independent precursor differentiation in the bone marrow, which is expressed on the surface membrane as the B-cell antigen receptor (BCR).

Andrew L. Snow and Michael J. Lenardo (eds.), *Immune Homeostasis: Methods and Protocols*, Methods in Molecular Biology, vol. 979, DOI 10.1007/978-1-62703-290-2_11,

Together, all newly produced B cells have a huge repertoire of antigen receptors with the potential to specifically recognize a vast array of different pathogens. The second phase of development starts when a B cell actually recognizes antigen in peripheral lymphoid organs. These activated B cells undergo enormous clonal proliferation, thereby generating huge numbers of daughter cells recognizing the same pathogen. This clonal expansion and selection of high affinity clones generates effector cells for a strong response and long-term memory in the form of memory B cells and Ig-producing plasma cells.

The host requires a highly dynamic immune system, which maintains a tight balance between the production of a large repertoire of B cells with unique receptors, and a strong immune response of groups of cells with an antigen-specific and thereby a more limited (selected) repertoire. In this regard, it is not surprising that the naive mature B cells are subject to multiple processes that regulate their dynamic homeostasis between output from bone marrow and antigen-dependent maturation (1, 2).

Recent bone marrow emigrants are functionally immature, i.e., they do not respond to BCR stimulation. In man, these transitional B cells constitute ~5–10% of total B cells in blood of healthy adults and have a characteristic phenotype that includes cell surface expression of CD10 as well as high expression of CD24 and CD38 (3, 4). In mouse, transitional B cells in spleen are $AA4.1^{+}$ (5, 6). Transitional B cells have not proliferated since their exit from bone marrow (7), and show no signs of selection for antibody reactivity as compared to immature B cells in bone marrow (8). In contrast, development from transitional B cells into $CD24^{dim}CD38^{dim}$ naive mature B cells in man and $AA4.1^{-}CD21^{dim}CD23^{hi}$ follicular B cells in mouse involves active selection processes (5, 8).

Furthermore, naive mature B cells critically depend on at least three survival signals (1). The first of these involves membrane expression of a BCR. In its absence, no B cells are generated, while conditional deletion in mature B cells induces cell death, indicating it is necessary for peripheral survival of B cells (9, 10). BCR-dependent survival signals are provided through a single pathway involving PI3K signaling (11), a conserved pathway for cellular homeostasis and survival (12).

In addition, naive mature B cells compete for two rate-limiting soluble factors for their survival. The production of the cytokine MIF (macrophage migration inhibitory factor) by dendritic cells in bone marrow is crucial for mature B-cell maintenance (13). MIF induces B-cell survival through binding to the CD74-CD44 receptor complex which subsequently stimulates hepatocyte growth factor (HGF)-induced c-MET signaling (14–16). While MIF is thought to act mainly on B cells migrating through bone marrow, soluble B-cell activating factor (BAFF) provides survival signals to naive mature B cells in peripheral lymphoid organs. Similar to a proliferation-inducing ligand (APRIL), BAFF is implicated in several aspects of B-cell survival as well as Ig isotype switching and production (17, 18). BAFF and APRIL can both bind the BCMA (B cell-maturation antigen) and TACI (transmembrane activator and CAML interactor) receptors (19–23), whereas BAFF also specifically binds to a third receptor, BAFF-R (24, 25), and APRIL can interact with proteoglycans (26, 27). BAFF-R is specifically expressed on B cells and the BAFF—BAFF-R interaction is crucial for survival of naive B cells (17, 28). Consequently, genetic ablation of BAFF or BAFF-R in mice and mutations in the human BAFF-R result in a dramatic reduction of naive mature B cells, while transitional B cells are hardly affected (29–31).

By using a real-time quantitative (RQ-)PCR-based method, we found that in contrast to transitional B cells, naive mature B cells in man and follicular B cells in mouse had undergone proliferation *in vivo* (7). This homeostatic proliferation, i.e., proliferation without further differentiation, was consistently found in healthy individuals to be ~2 cell divisions (7, 32, 33), while in some disease states, the replication history was increased to 6 or more cell divisions (34–36). The increased proliferation was associated with decreased bone marrow output, and is likely a mechanism to compensate for this. It remains unclear which factors drive this homeostatic proliferation, but the rate-limiting survival factors MIF and BAFF are likely candidates.

Our approach to measure proliferation exploits the circular DNA excision products that are generated in precursor B cells that undergo V(D)J recombination to generate a unique antibody. Most B cells initiate a single-step gene rearrangement to render a specific locus non-functional. This involves the rearrangement between the Jκ-Cκ intron recombination signal sequence (IRS1 in mouse and intronRSS in man) and the kappa-deleting element (RS in mouse and Kde in man) which deletes the constant region of 35–40% *IGK* alleles prior to migration of B cells to the periphery and is not followed by proliferation in bone marrow (Fig. 1.) (37). Thus, the coding joint to signal joint ratio of this rearrangement in a B-cell population can be used for robust quantification of replication history studies as (7): (a) it is a frequently occurring gene rearrangement; (b) it is one of the last Ig gene rearrangements to occur in bone marrow, ensuring that the corresponding kappa-deletion recombination excision circles (KRECs) are abundantly present in naive B lymphocytes; (c) it is a single-step rearrangement, which allows easy design of RQ-PCR primers and probes for accurate detection of the coding joints and signal joints; and (d) it is an end-stage rearrangement that deletes the enhancers, precluding further rearrangements.

In this chapter, we describe how to employ the KREC assay to study the replication history of purified human or mouse B-cell subsets.

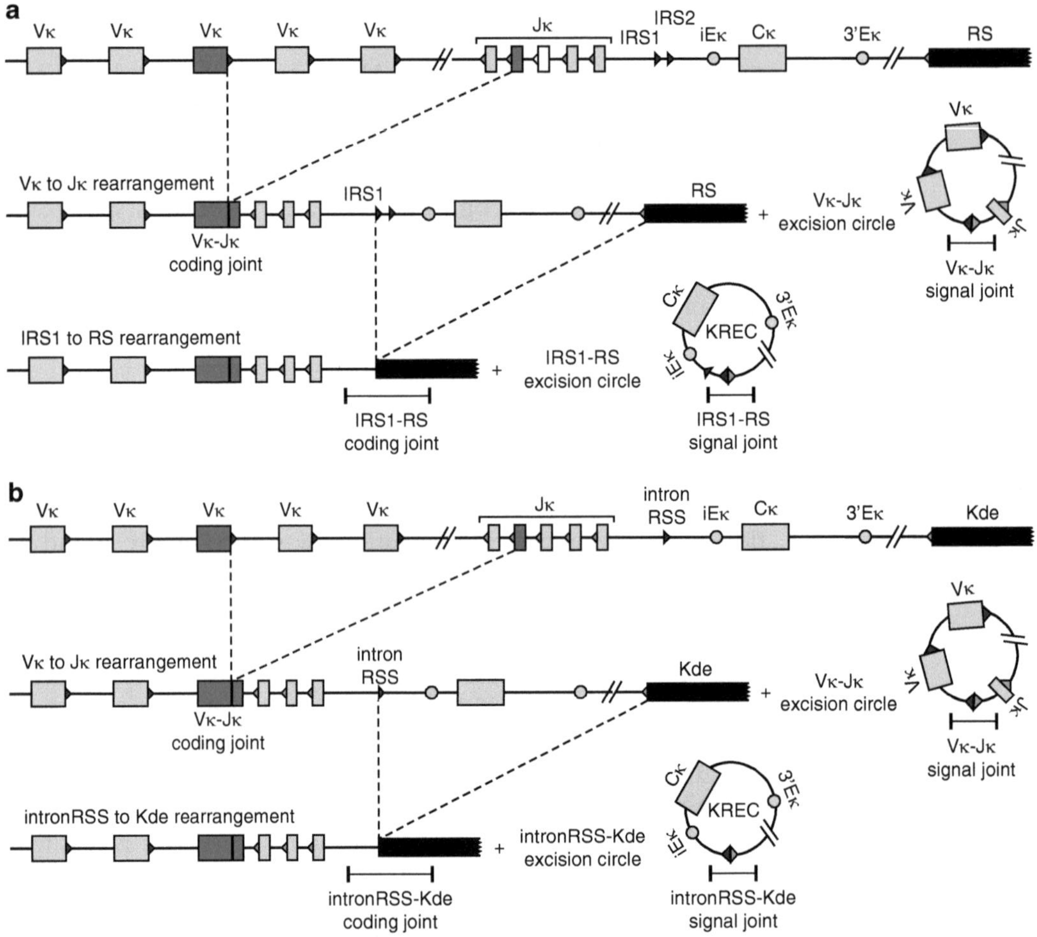

Fig. 1. Detection of coding joints and signal joints of kappa-deleting rearrangements in mice (**a**) and man (**b**). V(D)J recombination on the *IGK* locus results in a VJκ coding joint. Subsequent rearrangement between the IRS1 (mouse) or intronRSS (human) and the RS (mouse) and Kde (human) elements can make the *IGK* allele non-functional by deleting the Cκ exons and the enhancers. Consequently, the coding joint precludes any further rearrangements in the *IGK* locus and therefore remains present in the genome. The KREC with the corresponding signal joint is a stable double-stranded, circular DNA structure. The coding joint in the genome and the signal joint on the episomal excision circle can be quantified via real-time quantitative PCR using the indicated primers and TaqMan probes. The murine *IGK* locus contains two intronRSS sequences, IRS1 and IRS2, of which IRS1 has the most conserved RSS sequence and is ten times more frequently found in rearrangements to RS (M.C.v.Z., unpublished results) (41). Figure is reproduced from Van Zelm *et al.* 2007 (7) with permission from Rockefeller University Press.

2. Materials

2.1. High-Speed Cell Sorting

1. Phosphate-buffered saline (PBS), dissolve 8 g of NaCl, 0.2 g of KCl, 0.12 g of KH_2PO_4, and 1.67 g of $Na_2HPO_4{\cdot}2H_2O$ in approx. 800 ml of water, adjust pH to 7.8 with HCl, and add water to 1 L.

2. PBS/bovine serum albumin (BSA): PBS supplemented with 0.2% BSA and 0.1% azide.
3. Fluorochrome-conjugated antibodies (7, 33).
4. Fluorescence activated cell sorter (FACS), e.g., FACSAria (BD Biosciences).
5. Fetal Calf Serum (FCS).

2.2. DNA Isolation and TaqMan-Based Real-Time Quantitative (RQ-)PCR

1. Column-based miniprep DNA isolation kit, e.g., Sigma-Aldrich or Qiagen.
2. Spectrophotometer, e.g., NanoDrop.
3. Universal PCR Master Mix (Applied Biosystems), containing 5 mM $MgCl_2$, 200 μM dNTP, 0.1 U Taq Gold, and 10% glycerol.
4. 2% BSA.
5. Primers, in a 30 pmol/μl working dilution (Table 1).
6. 5′FAM and 3′TAMRA labeled hydrolysis probes, in a 5 pmol/μl working dilution (see Table 1. for specific oligonucleotide information).
7. Genomic DNA of the control cell line U698-DB01 that contains one intronRSS–Kde coding joint and one signal joint per genome (InVivoScribe; see Note 1) (7).
8. Plasmid controls for mouse IRS1–RS coding and signal joints. These can be constructed through ligation of PCR fragments in the pGEM®-T Easy vector (Promega; see Note 2) (7).
9. Thermocycler capable of detecting FAM signals from TaqMan-based hydrolysis probes, e.g., ABI Prism 7000 (Applied Biosystems).

3. Methods

3.1. High-Speed Cell Sorting

1. Start with mononuclear cells following Ficoll density separation (see Chapter 2), or with pre-sorted B cells, e.g., via $CD19^+$ enrichment with magnetic beads (Miltenyi Biotec).
2. Resuspend cell pellet after final wash in 500 μl PBS/BSA per 50×10^6 cells.
3. Add appropriate amounts of antibodies of choice, vortex briefly and incubate sample for 10 min at room temperature.
4. Wash sample once with 10–20× labeling volume of PBS/BSA, spin down at ~$300 \times g$ for 5–10 min.
5. Resuspend cells to a final concentration of 25×10^6 cells/ml in PBS/BSA.

Table 1
Oligonucleotides for quantification of coding joints and signal joints of kappa-deleting rearrangements

Target[a]	Forward primer sequence (5′–3′)	Reverse primer sequence (5′–3′)	TaqMan probe sequence (5′–3′)
Mouse studies			
IRS1-RS coding joint	CGCTAAGGGCCATGTGAAC	TTTAAAGCTACATTAGGGCTCAAATCT	TGGCAGCCCAGGGTGGATCTCC
IRS1-RS signal joint	GGAGTGGATTCAGGACACTGCT	CTCCAATAAGTCACCCTTTCCTTGT	CCAGTTTCTGCACGGGCAGTCAGTTAG
PGEMT-easy control	GTCACCTAAATAGCTTGGCGTAATC	CACGACAGGTTTCCCGACTG	CCACACAACATACGAGCCGGAAGCATAA
Human studies			
IntronRSS-Kde coding joint	CCCGATTAATGCTGCCGTAG	CCTAGGGAGCAGGGAGGCTT	AGCTGCATTTTTGCCATATCCACTATTTGGAGT
IntronRSS-Kde signal joint	TCAGCGCCCATTACGTTTCT	GTGAGGGACACGCAGCC	CCAGCTCTTACCCTAGAGTTTCTGCACGG
Human albumin[b]	TGAAACATACGTTCCCAAAGAGTTT	CTCTCCTTCTCAGAAAGTGTGCATAT	TGCTGAAACATTCACCTTCCATGCAGA

[a]See ref. (7)

[b]See refs. (39, 40)

6. Define B-cell populations (e.g., $CD19^{+}CD27^{-}IgM^{+}IgD^{+}$ naïve mature human B cells) with electronic gating on FACS and perform high-speed cell sorting.
7. Collect specified B-cell populations in tubes that contain a small volume (0.1–0.5 ml) of FCS.
8. Upon completion of cell sorting, spin cells down and start DNA isolation procedure.

3.2. DNA Isolation

1. Perform DNA isolation using a column-based kit according to the manufacturer's instructions (see Note 3).
2. Perform elution of DNA from column with elution buffer pre-heated to 55 °C. When starting with 30,000–200,000 cells, elute in 50 μl; 200,000–2 million cells in 100 μl; 2–5 million cells in 200 μl.
3. Measure DNA quantity and purity on a spectrophotometer.
4. If concentration is >15 ng/μl, prepare working dilution of 10 ng/μl in water.

3.3. TaqMan-Based Real-Time Quantitative (RQ-)PCR

1. For studies on human B cells, prepare three master mixes for independent RQ-PCR amplification of IntronRSS–Kde coding joints, signal joints and for the albumin control gene. Perform these three reactions in duplicate on each sample, on the control cell line U698-DB01 and on a non-template (water) control (see Note 4).
2. For studies on mouse B cells, prepare two master mixes for independent RQ-PCR amplification of IRS1–RS coding joints and signal joints. Perform these two reactions in duplicate on each sample, and on a non-template (water) control. Furthermore, perform independent amplification of the coding joint on the coding joint plasmid control, and of the signal joint on the signal joint plasmid control. In addition, perform the pGEMT vector backbone PCR on both plasmid constructs.
3. The master mix consists of 12.5 μl Universal PCR Master Mix, 0.75 μl of both the forward and the reverse primer (30 pmol/μl), 0.5 μl hydrolysis probe (5 pmol/μl), 0.5 μl 2% BSA, and 5 μl water per reaction.
4. Add 5 μl template DNA (working dilution 10 ng/μl) or water per reaction well.
5. Adjust the PCR program for FAM detection and TAMRA quenching. The thermocycler program consists of a 50 °C step for 2 min and 95 °C for 10 s, followed by 50× (95 °C for 15 s and 60 °C for 1 min).
6. Upon completion of RQ-PCR reaction, set the threshold for signal detection such that all samples are in the exponential

amplification phase. For consistency, it is preferred to set the same threshold for all three PCR reactions: intronRSS–Kde coding joint, signal joint and albumin for human studies, and IRS1–RS coding joint, signal joint and pGEMT vector control for mouse studies.

7. Extract the cycle threshold (Ct) value for each reaction and calculate the average Ct value for duplicate reactions. Duplicates should differ <1 Ct.

3.4. Calculation of Rearranged IntronRSS–Kde Alleles and the B-Cell Replication History

1. The ratio between the amount of genomic albumin and the intronRSS–Kde coding joints is indicative for the frequency of *IGK* alleles with an intronRSS–Kde rearrangement. Since the data are derived from independent PCRs, any technical variation is corrected for using the results on the U698-DB01 cell line which has one intronRSS–Kde coding joint per genome. The frequency of *IGK* alleles contain the intronRSS–Kde coding joint is thus calculated as follows (see Note 5)

$$2^{[(C_{T_{\text{albumin}}} - C_{T_{\text{coding joint}}})_{\text{sample}} - (C_{T_{\text{albumin}}} - C_{T_{\text{coding joint}}})_{\text{cell line}}]} .100\%$$

2. The difference in Ct values for the intronRSS–Kde coding joint and signal joints is directly related to the replication history of the B-cell subset (Fig. 2). Still, also for this calculation, the results of the U698-DB01 cell line have to be used to correct for differences in PCR efficiencies as follows:

$$(CT_{\text{signal joint}} - CT_{\text{coding joint}})_{\text{sample}} - (CT_{\text{signal joint}} - CT_{\text{coding joint}})_{\text{cell line}}$$

It is important to note that the increased proliferation will lead to reduced signal joints on KRECs (see Note 6).

3. While the calculation for the replication history of mouse B cell subsets is similar, the technical correction is slightly different because of the use of plasmid controls. The difference in efficiency of coding joint (cj) and signal joint (sj) PCRs of the IRS1–RS rearrangement can be corrected through the pGEMT PCR. Thus, the replication history is calculated as follows:

$$(CT_{\text{sj}} - CT_{\text{cj}})_{\text{sample}}((CT_{\text{sj}} - CT_{\text{pGEMT}})_{\text{sj construct}} - (CT_{\text{cj}} - C_{T\text{pGEMT}})_{\text{cj construct}})$$

4. Notes

1. The Igκ[+] B-cell line U698-M has two Vκ–Jκ rearranged *IGK* alleles, one of which is functional and has the intronRSS and Kde elements in germline configuration. The second allele has an out-of-frame VJκ joint and contains an intronRSS–Kde rearrangement. In order to obtain an equal ratio between the intronRSS–Kde

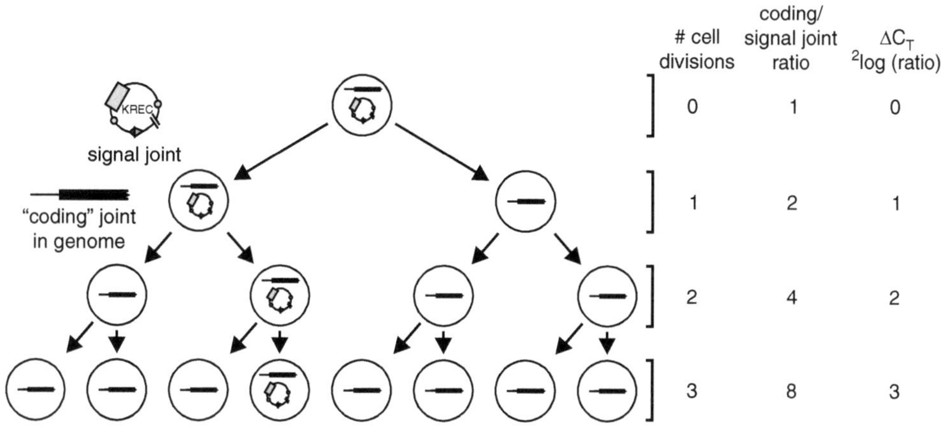

Fig. 2. Two-fold dilution of kappa-deleting recombination excision circles (KREC) with each cell division. When a B-lymphocyte with an intronRSS–Kde rearrangement divides, both daughter cells inherit the intronRSS–Kde coding joint in the genome. However, the signal joint, which is on the episomal KREC, will be inherited by only one of the two daughter cells. Crucially, the ΔCt of the PCRs detecting the coding joint and the signal joint exactly represent the number of cell divisions a B lymphocyte has undergone, because both processes have an exponential increase with base number 2. Figure is reproduced from Van Zelm *et al.* (7) with permission from Rockefeller University Press.

coding and signal joints, an intronRSS–Kde signal joint construct was inserted into the genome of the U698-M cell line using retroviral transduction. Individual clones were sorted, and the DB01 clone was found to contain a single insertion per genome, similar to the intronRSS–Kde coding joint (7). The primer-probe sets for both the coding and signal joint are optimized and show nearly equal efficiencies. Still, minimal variation (<1 Ct) can occur. Therefore, it is recommended to correct for this technical variation in each experiment.

2. Due to absence of a mouse control cell line with the IRS1–RS rearrangement, we developed an alternative approach to correct for technical variation between the coding joint and signal joint PCRs. Both coding joint and signal joint constructs were ligated into the pGEM®-T Easy vector to create independent controls on which the coding joint or signal joint PCR can be performed. When in parallel the pGEMT control PCR (Table 1) is run, the PCR efficiencies can be determined and compared.
3. It is recommended to perform the KREC assay on DNA isolated from at least 30,000 cells to ensure reliable signal joint amplification. When low replication histories are expected, it is still possible to obtain reliable signal joint amplification. However, for DNA isolation of low cell numbers, we recommend using a direct DNA lysis method (38).
4. For optimal calculation of the frequency of rearranged intronRSS–Kde alleles and B-cell replication history it is recommended to perform the coding joint, signal joint and albumin control PCRs in the same run. As with all RQ-PCR

experiments, we recommend performing each reaction at least in duplicate. In addition, each run should contain a non-template control and the U698-DB01 control for correction of technical variation (see Note 1).

5. In addition to the intronRSS–Kde coding to signal joint ratio, we recommend calculating the frequency of rearranged coding joints with the albumin control gene. While this is not crucial for the actual replication history levels, it provides valuable information on how representative this is for the whole population. In general, 35–50% of *IGK* alleles in mature B cells contain an intronRSS–Kde coding joint. While Igλ$^+$ B cells contain higher levels than Igκ$^+$ B cells (37), these rearrangements are independent of the antigen specificity. Thus, substantially high frequencies of intronRSS–Kde coding joints (>35%) reflect a good representation for the total B-cell subset.
6. With each cell division, the V(D)J recombination excision circles are diluted in the total B cell population. In memory B cell subsets of healthy controls, the replication history can be up to 10–12 cell divisions. This means that one circle is detected for 2^{10}–2^{12} (i.e., ~1,000–4,000) coding joints in the genome. Thus, when high replication histories are expected, it is important to start with high amounts of template DNA to maximize the quantitative range.

Acknowledgments

MAB is supported by a Fellowship from the Ter Meulen Fund–Royal Netherlands Academy of Arts and Sciences. MCvZ is supported by an Erasmus University Rotterdam (EUR)-Fellowship, an Erasmus MC Fellowship, by Veni grant 916.110.90 from ZonMW/NWO and by grant 689 of the Sophia Children's Hospital Fund.

Conflict of interest

JJMvD is inventor of the KREC assay, which has been patented (PCT/NL 2005/000761; priority date 25 Oct 2004) and licensed to InVivoScribe Technologies, San Diego, CA; revenues of the patent go to Erasmus MC. The other authors have declared that no conflict of interest exists.

References

1. Berkowska MA, van der Burg M, van Dongen JJM, van Zelm MC (2011) Checkpoints of B cell differentiation: visualizing Ig-centric processes. Ann N Y Acad Sci 1246:11–25
2. van Zelm MC, van der Burg M, van Dongen JJ (2007) Homeostatic and maturation-associated proliferation in the peripheral B-cell compartment. Cell Cycle 6:2890–2895
3. Cuss AK, Avery DT, Cannons JL, Yu LJ, Nichols KE, Shaw PJ, Tangye SG (2006) Expansion of functionally immature transitional B cells is associated with human-immunodeficient states characterized by impaired humoral immunity. J Immunol 176:1506–1516
4. Sims GP, Ettinger R, Shirota Y, Yarboro CH, Illei GG, Lipsky PE (2005) Identification and characterization of circulating human transitional B cells. Blood 105:4390–4398
5. Loder F, Mutschler B, Ray RJ, Paige CJ, Sideras P, Torres R, Lamers MC, Carsetti R (1999) B cell development in the spleen takes place in discrete steps and is determined by the quality of B cell receptor-derived signals. J Exp Med 190:75–89
6. Allman D, Lindsley RC, DeMuth W, Rudd K, Shinton SA, Hardy RR (2001) Resolution of three nonproliferative immature splenic B cell subsets reveals multiple selection points during peripheral B cell maturation. J Immunol 167:6834–6840
7. van Zelm MC, Szczepanski T, van der Burg M, van Dongen JJ (2007) Replication history of B lymphocytes reveals homeostatic proliferation and extensive antigen-induced B cell expansion. J Exp Med 204:645–655
8. Wardemann H, Yurasov S, Schaefer A, Young JW, Meffre E, Nussenzweig MC (2003) Predominant autoantibody production by early human B cell precursors. Science 301:1374–1377
9. Lam KP, Kuhn R, Rajewsky K (1997) In vivo ablation of surface immunoglobulin on mature B cells by inducible gene targeting results in rapid cell death. Cell 90:1073–1083
10. Torres RM, Flaswinkel H, Reth M, Rajewsky K (1996) Aberrant B cell development and immune response in mice with a compromised BCR complex. Science 272:1804–1808
11. Srinivasan L, Sasaki Y, Calado DP, Zhang B, Paik JH, DePinho RA, Kutok JL, Kearney JF, Otipoby KL, Rajewsky K (2009) PI3 kinase signals BCR-dependent mature B cell survival. Cell 139:573–586
12. Engelman JA, Luo J, Cantley LC (2006) The evolution of phosphatidylinositol 3-kinases as regulators of growth and metabolism. Nat Rev Genet 7:606–619
13. Sapoznikov A, Pewzner-Jung Y, Kalchenko V, Krauthgamer R, Shachar I, Jung S (2008) Perivascular clusters of dendritic cells provide critical survival signals to B cells in bone marrow niches. Nat Immunol 9:388–395
14. Gore Y, Starlets D, Maharshak N, Becker-Herman S, Kaneyuki U, Leng L, Bucala R, Shachar I (2008) Macrophage migration inhibitory factor induces B cell survival by activation of a CD74-CD44 receptor complex. J Biol Chem 283:2784–2792
15. Matza D, Lantner F, Bogoch Y, Flaishon L, Hershkoviz R, Shachar I (2002) Invariant chain induces B cell maturation in a process that is independent of its chaperonic activity. Proc Nat Acad Sci USA 99:3018–3023
16. Shachar I, Flavell RA (1996) Requirement for invariant chain in B cell maturation and function. Science 274:106–108
17. Mackay F, Schneider P, Rennert P, Browning J (2003) BAFF AND APRIL: a tutorial on B cell survival. Annu Rev Immunol 21:231–264
18. Dillon SR, Gross JA, Ansell SM, Novak AJ (2006) An APRIL to remember: novel TNF ligands as therapeutic targets. Nat Rev Drug Discov 5:235–246
19. Gross JA, Johnston J, Mudri S, Enselman R, Dillon SR, Madden K, Xu W, Parrish-Novak J, Foster D, Lofton-Day C, Moore M, Littau A, Grossman A, Haugen H, Foley K, Blumberg H, Harrison K, Kindsvogel W, Clegg CH (2000) TACI and BCMA are receptors for a TNF homologue implicated in B-cell autoimmune disease. Nature 404:995–999
20. Marsters SA, Yan M, Pitti RM, Haas PE, Dixit VM, Ashkenazi A (2000) Interaction of the TNF homologues BLyS and APRIL with the TNF receptor homologues BCMA and TACI. Curr Biol 10:785–788
21. Thompson JS, Schneider P, Kalled SL, Wang L, Lefevre EA, Cachero TG, MacKay F, Bixler SA, Zafari M, Liu ZY, Woodcock SA, Qian F, Batten M, Madry C, Richard Y, Benjamin CD, Browning JL, Tsapis A, Tschopp J, Ambrose C (2000) BAFF binds to the tumor necrosis factor receptor-like molecule B cell maturation antigen and is important for maintaining the peripheral B cell population. J Exp Med 192:129–135
22. Wu Y, Bressette D, Carrell JA, Kaufman T, Feng P, Taylor K, Gan Y, Cho YH, Garcia AD, Gollatz E, Dimke D, LaFleur D, Migone TS, Nardelli B, Wei P, Ruben SM, Ullrich SJ, Olsen HS, Kanakaraj P, Moore PA, Baker KP (2000) Tumor necrosis factor (TNF) receptor

superfamily member TACI is a high affinity receptor for TNF family members APRIL and BLyS. J Biol Chem 275:35478–35485
23. Yu G, Boone T, Delaney J, Hawkins N, Kelley M, Ramakrishnan M, McCabe S, Qiu WR, Kornuc M, Xia XZ, Guo J, Stolina M, Boyle WJ, Sarosi I, Hsu H, Senaldi G, Theill LE (2000) APRIL and TALL-I and receptors BCMA and TACI: system for regulating humoral immunity. Nat Immunol 1:252–256
24. Thompson JS, Bixler SA, Qian F, Vora K, Scott ML, Cachero TG, Hession C, Schneider P, Sizing ID, Mullen C, Strauch K, Zafari M, Benjamin CD, Tschopp J, Browning JL, Ambrose C (2001) BAFF-R, a newly identified TNF receptor that specifically interacts with BAFF. Science 293:2108–2111
25. Yan M, Brady JR, Chan B, Lee WP, Hsu B, Harless S, Cancro M, Grewal IS, Dixit VM (2001) Identification of a novel receptor for B lymphocyte stimulator that is mutated in a mouse strain with severe B cell deficiency. Curr Biol 11:1547–1552
26. Hendriks J, Planelles L, de Jong-Odding J, Hardenberg G, Pals ST, Hahne M, Spaargaren M, Medema JP (2005) Heparan sulfate proteoglycan binding promotes APRIL-induced tumor cell proliferation. Cell Death Differ 12:637–648
27. Ingold K, Zumsteg A, Tardivel A, Huard B, Steiner QG, Cachero TG, Qiang F, Gorelik L, Kalled SL, Acha-Orbea H, Rennert PD, Tschopp J, Schneider P (2005) Identification of proteoglycans as the APRIL-specific binding partners. J Exp Med 201:1375–1383
28. Schneider P (2005) The role of APRIL and BAFF in lymphocyte activation. Curr Opin Immunol 17:282–289
29. Gross JA, Dillon SR, Mudri S, Johnston J, Littau A, Roque R, Rixon M, Schou O, Foley KP, Haugen H, McMillen S, Waggie K, Schreckhise RW, Shoemaker K, Vu T, Moore M, Grossman A, Clegg CH (2001) TACI-Ig neutralizes molecules critical for B cell development and autoimmune disease. impaired B cell maturation in mice lacking BLyS. Immunity 15:289–302
30. Schiemann B, Gommerman JL, Vora K, Cachero TG, Shulga-Morskaya S, Dobles M, Frew E, Scott ML (2001) An essential role for BAFF in the normal development of B cells through a BCMA-independent pathway. Science 293:2111–2114
31. Warnatz K, Salzer U, Rizzi M, Fischer B, Gutenberger S, Bohm J, Kienzler AK, Pan-Hammarstrom Q, Hammarstrom L, Rakhmanov M, Schlesier M, Grimbacher B, Peter HH, Eibel H (2009) B-cell activating factor receptor deficiency is associated with an adult-onset antibody deficiency syndrome in humans. Proc Nat Acad Sci USA 106:13945–13950
32. Moir S, Ho J, Malaspina A, Wang W, DiPoto AC, O'Shea MA, Roby G, Kottilil S, Arthos J, Proschan MA, Chun TW, Fauci AS (2008) Evidence for HIV-associated B cell exhaustion in a dysfunctional memory B cell compartment in HIV-infected viremic individuals. J Exp Med 205:1797–1805
33. Berkowska MA, Driessen GJ, Bikos V, Grosserichter-Wagener C, Stamatopoulos K, Cerutti A, He B, Biermann K, Lange JF, van der Burg M, van Dongen JJ, van Zelm MC (2011) Human memory B cells originate from three distinct germinal center-dependent and -independent maturation pathways. Blood 118:2150–2158
34. Rakhmanov M, Keller B, Gutenberger S, Foerster C, Hoenig M, Driessen G, van der Burg M, van Dongen JJ, Wiech E, Visentini M, Quinti I, Prasse A, Voelxen N, Salzer U, Goldacker S, Fisch P, Eibel H, Schwarz K, Peter HH, Warnatz K (2009) Circulating CD21low B cells in common variable immunodeficiency resemble tissue homing, innate-like B cells. Proc Nat Acad Sci USA 106:13451–13456
35. van der Burg M, Pac M, Berkowska MA, Goryluk-Kozakiewicz B, Wakulinska A, Dembowska-Baginska B, Gregorek H, Barendregt BH, Krajewska-Walasek M, Bernatowska E, van Dongen JJ, Chrzanowska KH, Langerak AW (2010) Loss of juxtaposition of RAG-induced immunoglobulin DNA ends is implicated in the precursor B-cell differentiation defect in NBS patients. Blood 115:4770–4777
36. Driessen GJ, van Zelm MC, van Hagen PM, Hartwig NG, Trip M, Warris A, de Vries E, Barendregt BH, Pico I, Hop W, van Dongen JJ, van der Burg M (2011) B-cell replication history and somatic hypermutation status identify distinct pathophysiological backgrounds in common variable immunodeficiency. Blood 118:6814–6823
37. van Zelm MC, van der Burg M, de Ridder D, Barendregt BH, de Haas EF, Reinders MJ, Lankester AC, Revesz T, Staal FJ, van Dongen JJ (2005) Ig gene rearrangement steps are initiated in early human precursor B cell subsets and correlate with specific transcription factor expression. J Immunol 175:5912–5922
38. van der Burg M, Kreyenberg H, Willasch A, Barendregt BH, Preuner S, Watzinger F, Lion T, Roosnek E, Harvey J, Alcoceba M, Diaz MG, Bader P, van Dongen JJ (2011) Standardization of DNA isolation from low cell numbers for chimerism analysis by PCR of short tandem repeats. Leukemia 25: 1467–1470

39. Pongers-Willemse MJ, Verhagen OJ, Tibbe GJ, Wijkhuijs AJ, de Haas V, Roovers E, van der Schoot CE, van Dongen JJ (1998) Real-time quantitative PCR for the detection of minimal residual disease in acute lymphoblastic leukemia using junctional region specific TaqMan probes. Leukemia 12:2006–2014
40. Van Zelm MC, Van Der Burg M, Langerak AW, Van Dongen JJM (2011) PID comes full circle: applications of V(D)J recombination excision circles in research, diagnostics and newborn screening of primary immunodeficiency disorders. Front Immun 2:1–9
41. Shimizu T, Iwasato T, Yamagishi H (1991) Deletions of immunoglobulin C kappa region characterized by the circular excision products in mouse splenocytes. J Exp Med 173:1065–1072

Chapter 12

Molecular Measurement of T Cell Receptor Excision Circles

Heather E. Lynch and Gregory D. Sempowski

Abstract

This chapter provides protocols necessary for quantifying human, mouse, and nonhuman primate signal joint T cell receptor excision circles (sjTRECs) produced during T cell receptor alpha (TCRA) gene rearrangement. These non-replicated episomal circles of DNA are generated by the recombination process used to produce antigen-specific T cell receptors. The number of sjTRECs per mg of thymus tissue or per 100,000 lysed cells has been shown to be a molecular marker of thymopoiesis and naïve T cells. This technology is beneficial to investigators interested in quantitating the level of naïve T cell production occurring under various circumstances in a variety of systems, and complements traditional phenotypic analyses of thymopoiesis. This chapter specifically describes procedures required for rapid detection and quantitation of sjTRECs in thymus tissue or isolated cells using real-time quantitative polymerase chain reaction (PCR). The sjTREC assay system comprises species-specific forward and reverse primers for amplification of a unique site on the T cell receptor δ (*TCRD*) sjTREC, a fluorescently labeled (FAM/ZEN/IABkFQ) species-specific real-time probe, and a species-specific sjTREC DNA plasmid standard for quantitation.

Key words: Human, Mouse, Non human Primate, Thymopoiesis, T cell receptor, Real-time PCR

1. Introduction

The peripheral T cell pool is established early in fetal development by education of bone marrow-derived T cell progenitors on the thymic stroma, with subsequent emigration of mature naïve T cells to peripheral sites such as lymph nodes, gut lymphoid tissue, and spleen (1). This process of thymopoiesis is essential for establishing the peripheral T cell pool early in life, and it has recently been shown that thymic output occurs well into the fourth and fifth decades of life (2–5).

A breakthrough in the study of thymopoiesis and peripheral T cell homeostasis in humans was the development of the T cell receptor excision circle (TREC) assay for the study of thymic function in vivo (6, 7). Kong et al. (7) showed that excised T cell receptor DNA circles were present in recently produced T cells and that

Andrew L. Snow and Michael J. Lenardo (eds.), *Immune Homeostasis: Methods and Protocols*, Methods in Molecular Biology, vol. 979, DOI 10.1007/978-1-62703-290-2_12,

these extrachromosomal DNA circles are byproducts of TCR gene rearrangement that are not replicated and diluted by T cell proliferation. Douek et al. (6, 8) developed a real-time PCR assay for the quantitation of signal joint T cell receptor excision circles (sjTRECs) in humans and reported that they localize in naïve-phenotype T cells and that their frequency decreases in peripheral blood $CD4^+$ and $CD8^+$ T cells with increasing age. The number of sjTRECs per mg of thymus tissue or per 100,000 lysed cells is used as a molecular marker of thymopoiesis and naïve T cells. Thus, measurement of sjTRECs has provided an invaluable assay for rapid assessment of thymic function and the status of T cell immune reconstitution in humans (6, 9–12).

The widespread use of mouse models in the study of T cell biology and immune reconstitution led to the development of a mouse sjTREC PCR assay by our group, based on the human assay (13). The TCRA gene rearrangement event monitored by the human sjTREC assay is the generation of an unique signal joint between δRec and Jα on extrachromosomal circles of DNA in T cells (8, 14). These pseudogenes are two genetic elements of V and Jα, respectively, that are conserved between humans and mice (15). The BALB/c mouse homolog of Jα is 3.1 kb upstream of the most upstream Jα (16). However, three reported murine δRec homologs exist that can be utilized to generate murine TCRD excision circles with the Jα. δRec1 and δRec2 have been identified in the region upstream of Dδ2 (17), and a putative δRec3 has been described 1.6 kb 3′ of Dδ1 (18). The assay described here detects the unique signal joint formed between the murine Jα and δRec1 because these elements were determined to rearrange at a high frequency in mouse thymus (14). In addition, the mouse primers and probes were designed using the BALB/c mouse TCRA/D gene sequence (Gen-Bank AE008686). Although the assay can be used for other strains, it is optimal for BALB/c mice.

The mouse sjTREC PCR assay has been used in a variety of experimental settings. It has been used to characterize murine thymic function during aging, as well as to study the effect of IL-7 on thymopoiesis and peripheral T cell expansion in young and aged mice (13). Mouse sjTREC PCR analysis has also been used to monitor thymopoiesis in neonates (19), investigate murine models of stem cell transplantation (20), and to measure recovery of thymopoiesis following stress-induced thymic atrophy (21–24). Factors that can influence peripheral sjTREC levels are thymic output, T cell proliferation (either homeostatic or antigen-driven), and trafficking in and out of lymphoid tissues.

As was done with mice, Sodora et al. adapted the human sjTREC assay to monitor thymopoiesis in nonhuman primates (25). Measurements of sjTREC were specifically used to show that, as in humans, thymopoiesis decreases across the lifespan of nonhuman primates (rhesus macaques and sooty mangabeys) (25). The nonhuman primate sjTREC assay has been used to assess thymopoiesis in these species and in cynomolgus monkeys following simian immuno deficiency virus infection and/or interleukin-7 treatment (26–29). Primer and probe sequences are provided in this chapter for sjTREC measurement in rhesus, mangabey, and cynomolgus systems.

This chapter describes a generalized real-time PCR assay for the rapid detection and quantitation of human, mouse, or nonhuman primate sjTRECs using two readily available real-time thermal cyclers: Bio-Rad iCycler iQ and Bio-Rad CFX 96 Real-Time System. The sjTREC assay system comprises species-specific forward and reverse PCR primers for amplification of an unique site on the T cell receptor δ (*TCRD*) sjTREC, a fluorescently labeled (FAM/ZEN/IABkFQ) species-specific real-time PCR probe for *TCRD* sjTREC, and a species-specific sjTREC DNA plasmid standard for quantitation (i.e., human, mouse, nonhuman primate). The DNA standard (calibrated in number of molecules) and samples are amplified for 45 PCR cycles and quantitated using a real-time thermal cycler. As *Taq* DNA polymerase amplifies the unique *TCRD* sjTREC sequence, annealed quiescent probe is digested by the nuclease activity of *Taq* and FAM fluorescence is liberated. Fluorescence is detected at each cycle and used to calculate molecules of sjTREC using a standard curve generated by the system software. Here we describe procedures for the preparation of samples to be assayed, either genomic DNA extracted from thymus tissue or proteinase K-digested lysates of isolated cells, and for the preparation of working stocks of the sjTREC DNA plasmid standards. For laboratories setting up the sjTREC PCR assay for the first time, it is important that three to 5 days be devoted to growing up the sjTREC DNA standard plasmid(s) and freezing down a large supply of pre-diluted aliquots of the standards. This initial investment will save time in the future and generate sufficient aliquots of standards for 2–3 years worth of assays.

2. Materials

The single most critical parameter when performing quantitative real-time PCR is to avoid contamination of the work area, equipment, reagents, and samples. All reagents should be prepared as described, observing stringent molecular biology technique. Purchase molecular biology-grade stock reagents and use aerosol-resistant pipet tips for all procedures. Wear gloves for all reagent and buffer preparation and change gloves regularly.

2.1. Miscellaneous Supplies

1. Pre-weighed thymus tissue biopsy or pre-counted and pelleted cell preparations (i.e., thymocytes or peripheral T cell subsets) (see Note 1).

2. Sterile forceps.
3. Trizol Reagent (Invitrogen), room temperature.
4. Soft tissue homogenizing CK14 tubes (1.4 mm ceramic beads in 2 mL tubes; Bertin Technologies).
5. Molecular-biology grade chloroform.
6. 100% and 75% molecular-biology grade ethanol.
7. Sodium citrate at 0.1 M in 10% ethanol.
8. DNase/RNase-free molecular-biology grade water.
9. 1.5 mL polypropylene screw-cap tubes (Sarstedt brand suggested).
10. 15 mL polypropylene conical tubes.
11. 10 mM Tris–Cl, pH 7.8 in DNase/RNase-free H_2O.
12. 19.2 mg/mL proteinase K (PCR Grade, Roche).
13. Yeast tRNA (PCR Grade).

2.2. Species-Specific sjTREC Plasmid Standard

1. Grow up a stock of human, mouse, or rhesus sjTREC plasmid (Gregory D. Sempowski, Duke University; gregory. sempowski@duke.edu) using standard molecular biology techniques. Determine the concentration in μg/mL of the purified plasmid (see Note 2).
2. Using the species-specific plasmid molecular weight, determine the number of molecules of sjTREC plasmid per μL in the plasmid standard preparation (see Note 3).
3. Prepare a stock of 2×10^{10} sjTREC molecules/μL in DNase/RNase-free water.
4. Make serial dilutions of 100 μL standard into 900 μl DNase/RNase-free water for 10^{10}, 10^{9}, and 10^{8} molecules per 5 μL aliquot. Vortex well to mix and microcentrifuge briefly at maximum speed before opening tubes.
5. For standard dilutions of 10^{7}, 10^{6}, 10^{5}, 10^{4}, 10^{3}, and 10^{2} molecules/5 μL, dilute 500 μL of the 10^{8} dilution up to 5 mL in a 15 mL polypropylene conical tube with water containing 30 ng/mL yeast tRNA (see Note 4). Vortex well to mix and centrifuge briefly at maximum speed in a tabletop centrifuge before opening tubes.
6. Validate prepared standard dilutions by running 5 μL of each dilution (10^{7} through 10^{2}) in the sjTREC PCR assay (see Note 5).
7. Aliquot each standard at 1 mL/tube, but aliquot the last milliliter of each standard at 200 μL/tube. Each 200 μL tube is aliquoted as 15 μL/tube working aliquots as needed. One 15 μL aliquot of each standard will be needed per sjTREC PCR assay run.

8. Freeze all standard stocks and dilutions immediately at −80°C. Store standards in a separate box away from all other PCR reagents and experimental samples. Label all tubes well and avoid freeze/thaw of the working aliquots of standards.

2.3. PCR Reaction

1. Platinum Taq DNA polymerase, 50 mM $MgCl_2$, and reaction buffer (Invitrogen or equivalent).
2. 10 mM dNTP mix: To prepare dNTP mix, combine 1 M stock solutions of dATP, dCTP, dGTP, and dUTP at a ratio of 1:100 in DNase/RNase-free water such that each deoxynucleotide is present in the mixture at a final concentration of 10 mM. Mix solution well by vortexing and store up to 2 years at −80°C in 500 μL aliquots.
3. 96-well PCR plates and PCR plate strip caps (see Note 6).

2.4. Species-Specific Primers

1. 12.5 μM stocks of species-specific sjTREC primers (see Notes 7–9).
2. Reconstitute lyophilized primers with water to a stock concentration of 125 μM. Store 50 μL aliquots at −80°C.
3. Prepare a working stock by diluting 1:10 with water to a final concentration of 12.5 μM, aliquot at 200 μL/tube, and store at −80°C.

2.5. Species-Specific Probes

1. 5 μM Stock of species-specific sjTREC probe (see Notes 7–9).
2. Real-time PCR probes (Integrated DNA Technologies) contain a 5′ reporter fluorochrome (FAM), an internal quencher (ZEN), and a 3′ dark quencher (IABkFQ). This double-quenching design allows for lower background and higher signal than single-quench probes.
3. Reconstitute lyophilized probe with water to a stock concentration of 50 μM. Store 50 μL aliquots at −80°C.
4. Prepare a working stock by diluting 1:10 with water to a final concentration of 5 μM, aliquot at 200 μL/tube, and store at −80°C.

2.6. Equipment

1. 55°C water bath.
2. Tabletop centrifuge with swinging 96-well plate holders and 15 mL tube holders.
3. Microcentrifuge.
4. Thermomixer (Eppendorf).
5. Precellys-24 Homogenizer (Bertin Technologies; or similar).
6. PCR setup hood (with optional UV lamp to reduce the possibility of background contamination).

7. Bio-Rad iCycler iQ Thermal Cycler or Bio Rad CFX96 Real-Time System (or similar) with optical system and filter sets for detection of FAM.

3. Methods

To avoid contamination of the work area, equipment, reagents, and samples all reagents should be prepared as described, observing stringent molecular biology technique. Use aerosol-resistant pipet tips for all procedures. Wear gloves for all procedures and change gloves when transitioning between setting up the PCR reaction mix, the addition of the samples, and the addition of the standards.

3.1. Preparation of Thymus Tissue Genomic DNA

1. Using sterile forceps, transfer frozen tissue biopsy to ceramic bead tube containing 1 mL Trizol reagent.
2. Load tubes into Precellys-24 homogenizer and process for 20 s at 5,000 rpm.
3. Using a micropipettor with a 200 μL tip, transfer tissue homogenate into a pre-labeled 1.5 mL microcentrifuge tube.
4. Add 200 μL of chloroform to each microcentrifuge tube containing the 1 mL homogenized tissue/Trizol. Cap tubes well and vortex for 15 s. Incubate at room temperature (RT) for 15 min.
5. Microcentrifuge samples 15 min at ~13,000 × *g*, 2–8°C. After centrifugation, the aqueous phase will contain RNA, the interphase will contain cellular proteins and some DNA, and the organic (pink) phase will contain genomic DNA.
6. Carefully remove aqueous phase using a micropipettor with a 200 μL tip. Aqueous phase can be stored in a cryovial at −80°C if future extraction of RNA is desired.
7. Add 300 μL of 100% ethanol to the interphase/organic phase remaining in the microcentrifuge tube. Vortex gently. Incubate samples 2–3 min at RT.
8. Microcentrifuge 5 min at ~13,000 × *g* at RT.
9. Remove Trizol/ethanol supernatant to an appropriate waste container.
10. Add 1 mL 0.1 M sodium citrate/10% ethanol to each sample tube. Incubate samples 30 min at 15–30°C with periodic mixing.
11. Microcentrifuge 5 min at ~13,000 × *g*, room temperature. Repeat 0.1 M sodium citrate wash as in step 10.

12. Precipitate the DNA in 1 mL 75% ethanol. Incubate 10–20 min at 15–30°C with periodic mixing.
13. Microcentrifuge 5 min at 13,000 xg, room temperature. Remove supernatant using a micropipettor with a 200 μL tip.
14. Briefly dry pellet (no more than 5 min) under vacuum in a Speedvac evaporator or air-dry in a fume hood. Do not dry completely or the DNA will be difficult to resuspend. Dissolve pellet in 200–500 μL water. Record the precise volume used to dissolve the pellet. If DNA does not dissolve, place samples at 55°C for 10 min to increase solubility. Cool on ice, vortex, microcentrifuge briefly at maximum speed to collect solution at bottom of tube, and transfer solubilized DNA to a fresh 1.5 mL microcentrifuge tube.
15. Quantitate DNA using a UV spectrophotometer (read absorbance at 260 nm). Multiply A260 (OD) × dilution × 50 (μg/mL/OD) to determine DNA concentration in μg/mL. Multiply concentration by the total volume of DNA to determine total DNA yield. Divide the total μg of DNA by the initial weight of tissue used to determine μg of DNA per milligram of thymus tissue.
16. Freeze DNA samples at −80°C or proceed to Subheading 3.3.

3.2. Proteinase K Lysis of Isolated Lymphocytes

1. Calculate volume of proteinase K needed per sample by multiplying the total cell count by 0.0001; the result is the amount of proteinase K working solution in μL to add to the pellet for a final concentration of 10,000 cells/μL proteinase K.
2. Immediately before use, dilute the 19.2 mg/mL stock solution of proteinase K 1:200 with 10 mM Tris–HCl, pH 7.8. Layer this solution (using the volume calculated in step 1) on top of each pellet without letting the pipet tip touch the pellet. Vortex the tube and flick down to get the liquid in the bottom of the tube.
3. Place tubes in a Thermomixer at 56°C and shake for 1 h at 1,200 rpm.
4. Turn Thermomixer up to 95°C for 10 min to inactivate the proteinase K. Vortex tubes and microcentrifuge 1 min at ~13,000 × *g*.
5. Store lysates at −80°C or proceed to Subheading 3.3.

3.3. Quantitation of TCR Delta Excision Circles by Real-Time PCR Using the Bio-Rad iCycler IQ or CFX96 Real-Time Thermal Cycler Machines

1. Design the plate layout for the sjTREC PCR assay (Fig. 1). Determine the number of samples, standards, and no-template controls (NTC) to be run in duplicate. Add two extra wells to the total number of wells needed to allow for pipetting error.
2. Prepare the PCR reaction master mix (see Note 10). For each of the wells to be included in the assay (as determined in step

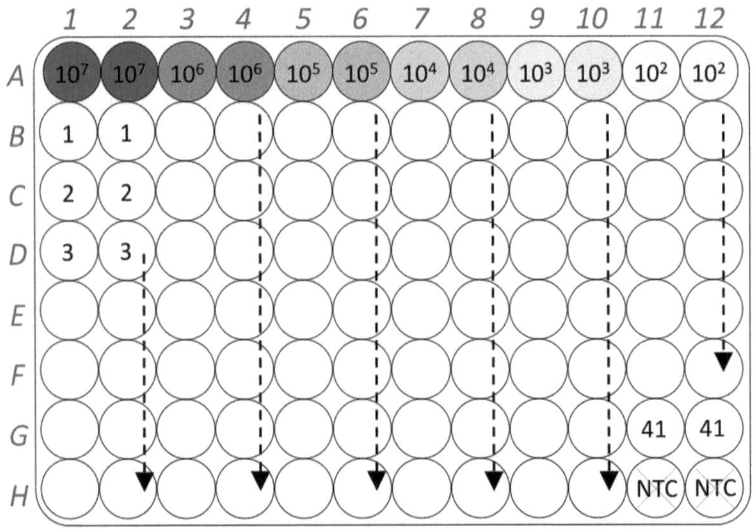

Fig. 1. Suggested layout of duplicate sjTREC standards (10^7 to 10^2 molecules), no-template controls (NTC), and experimental samples (1–41) on a 96-well PCR plate.

1), add the following to a 1.5 mL tube to prepare a PCR master mix:

12.125 μL H_2O

2.5 μL Platinum *Taq* buffer

1.75 μL 50 mM $MgCl_2$

0.5 μL 10 mM dNTP mix

1.0 μL forward primer working solution

1.0 μL reverse primer working solution

1.0 μL FAM probe

0.125 μL Platinum *Taq* enzyme

3. Vortex gently to mix well, then microcentrifuge briefly at maximum speed to collect the solution at the bottom of the tube.
4. Add 20 μL of PCR reaction mix to each well according to the plate layout (see step 1 and Fig. 1).
5. Add 5 μL of DNase/RNase-free water to each no-template control (NTC) well, according to the plate layout, and cap the wells.
6. Prepare DNA samples by diluting 3 μg of each sample to a final volume of 15 μL with water. Use 5 μL (1 μg DNA) of diluted DNA per well in the sjTREC PCR assay. Add 5 μL of each experimental sample (either 1 μg of DNA in 5 μL water or the volume equivalent (5 μL) of 50,000 cells of a proteinase K cell

lysate) into duplicate wells according to the plate layout. Cap wells after each row.

7. In a separate clean area, add 5 μL of each pre-diluted standard (10^2–10^7 molecules/5 μL) to duplicate wells according to the plate layout. Cap wells (see Note 11).
8. Gently vortex plate and centrifuge 5 min at ~400 × *g* in a tabletop centrifuge with 96-well plate holders.
9. Program the BioRad iCycler or CFX96 system for a 25 μL sample as follows:

   Cycle 1: (1 time) step 1, 95°C for 10 min

   Cycle 2: (45 times) step 1, 95°C for 15 s; step 2, 60°C for 1 min

   Cycle 3: (1 time) step 1, 4°C HOLD

10. Enable real-time data collection of FAM signal during Cycle 2 and use the "heated lid" option. Place plate in the thermal cycler, lock lid, and start run (see Note 12).
11. When the run is complete, remove the 96-well plate from the thermal cycler and discard.
12. Set the threshold at the midpoint of the linear amplification range of the standards (see system software manual). Analyze the r^2 and *y* intercept of the standard curve (see Note 13).

4. Notes

1. Fresh thymus tissue samples (~100 mg) should be snap-frozen in a dry ice/ethanol bath and stored at −80°C or in liquid nitrogen until ready for TREC assay. Isolated cell samples (250,000 cell minimum) should be thawed at 37°C, washed with 10 mL PBS, and then pelleted by centrifugation in a tabletop centrifuge at ~400 × *g* for 5 min at 4°C.
2. The biggest source of contamination is the plasmid that contains the mouse sjTREC DNA standard. Avoid working with the plasmid or aliquots of diluted standard when containers of other assay reagents or experimental samples are open.
3. The weight of one molecule of the sjTREC plasmids are as follows: mouse = 2.25×10^{-18} g; human = $4.19\text{x}10^{-18}$ g; nonhuman primate (rhesus) = 3.37×10^{-18} g.
4. Yeast tRNA is required to stabilize the diluted plasmid.
5. The Ct values for the standards should be evenly spaced over the 5-log curve. Ten-fold dilutions read out with a 3 cycle difference in Ct. The *y* intercept of the curve should be 45 ± 3, and have an $r^2 > 0.995$. Quantitated and prediluted aliquots of

sjTREC plasmid DNA are available for calibration purposes (Gregory D. Sempowski, Duke University).

6. It is important to use plates and caps designed for the specific thermal cycler being used, as these machines are calibrated for the specific material, density, color, and refractive angle of the plasticware.
7. Mouse forward sjTREC primer: 5′-CAT TGC CTT TGA ACC AAG CTG-3′. Mouse reverse sjTREC primer: 5′-TTA TGC ACA GGG TGC AGG TG-3′. Mouse probe: 5′-/56-FAM/ CA GGG CAG G/ZEN/T TTT TGT AAA GGT GCT CAC TT/3IABkFQ/-3′.
8. Human forward sjTREC primer: 5′-CAC ATC CCT TTC AAC CAT GCT-3′. Human reverse sjTREC primer: 5′-GCC AGC TGC AGG GTT TAG G-3′. Human probe: 5′-/56-FAM/AC ACC TCT G/ZEN/G TTT TTG TAA AGG TGC CCA CT/3IABkFQ/-3′.
9. Nonhuman primate (Rhesus, NHP) forward sjTREC primer: 5′-CAC ATC CCT TTC AAC CAT GCT-3′. NHP reverse sjTREC primer: 5′-GCC AGC TGC AGG GTT TAG G-3′. NHP Probe: 5′-/56-FAM/AC GCC TCT G/ZEN/G TTT TTG TAA AGG TGC TCA CT/3IABkFQ/-3′.
10. Prepare PCR reaction mix and add to the plate in a PCR hood or other ultraclean PCR preparation area. Do not, under any circumstances, use sjTREC standards in this setup area.
11. It is very important to have a separate work area for the sjTREC standards. The plasmid is highly concentrated, and contamination of experimental samples or PCR master mix reagents is likely unless separate work areas are established.
12. Refer to the Bio-Rad iCycler iQ Real-Time PCR Detection System or CFX Manager software user's manual for specifics on programming and operation of the iCycler or CFX96 real-time thermal cyclers.
13. Raw fluorescence data will be presented as Ct values, i.e., the PCR cycle at which the FAM reporter signal crosses the threshold setting. System software calculates Ct values of experimental samples and then converts them to number of molecules of sjTREC by comparing sample Ct to the standard curve. No template control wells should have a Ct value of 45 (negative). Sample values will be reported as number of sjTREC molecules per μg of DNA, or number of sjTREC molecules per 50,000 cells. Results (sjTREC/μg DNA or sjTREC/50,000 cells) will vary depending on the experimental samples. Typical sjTREC levels in normal BALB/c thymus tissue and isolated $CD4^+$ and $CD8^+$ splenocytes throughout aging are detailed in the initial publication of the mouse

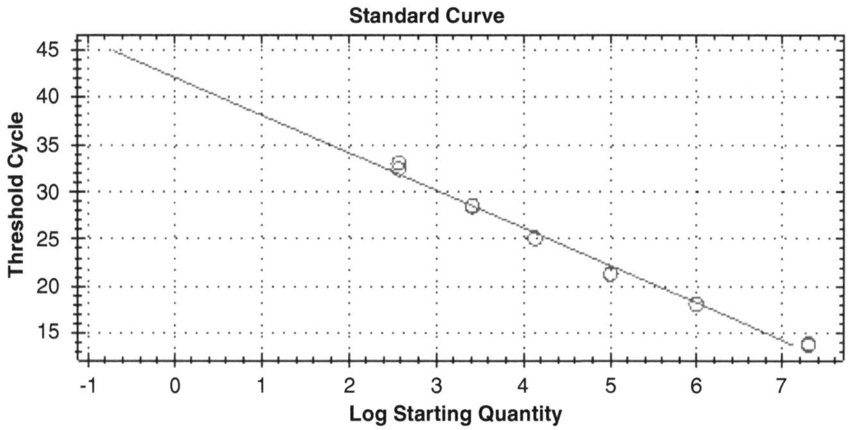

Fig. 2. Representative mouse sjTREC standard curve generated using the CFX96 Real-Time System.

sjTREC PCR assay (13). The iCycler iQ and the CFX96 systems will give similar results. The linear range of detection of human, mouse, and NHP sjTREC in 1 μg DNA or 50,000 cells is 10,000,000 to 100 molecules. A representative standard curve from the CFX96 Real-Time System is shown in Fig. 2. Typical runs on the BioRad systems use auto-calculated baseline determination, and the threshold value is generally placed between 17 and 28. PCR efficiency should be >80%. If these criteria are not met, then the run should be repeated with fresh aliquots of all reagents.

Acknowledgments

This work was supported by NIH grants AG-025150 and AI-067798 and was performed in the Regional Biocontainment Laboratory at Duke which received partial support for construction from NIH/NIAID (UC6-AI-058607).

References

1. Haynes BF, Denning SM, Le PT, Singer KH (1990) Human intrathymic T cell differentiation. Semin Immunol 2:67–77
2. Flores KG, Li J, Sempowski GD, Haynes BF, Hale LP (1999) Analysis of the human thymic perivascular space during aging. J Clin Invest 104:1031–1039
3. Jamieson BD, Douek DC, Killian S, Hultin LE, Scripture-Adams DD, Giorgi JV, Marelli D, Koup RA, Zack JA (1999) Generation of functional thymocytes in the human adult. Immunity 10:569–575
4. Sempowski GD, Hale LP, Sundy JS, Massey JM, Koup RA, Douek DC, Patel DD, Haynes

BF (2000) Leukemia inhibitory factor, oncostatin M, IL-6, and stem cell factor mRNA expression in human thymus increases with age and is associated with thymic atrophy. J Immunol 164:2180–2187
5. Lynch HE, Goldberg GL, Chidgey A, Van den Brink MR, Boyd R, Sempowski GD (2009) Thymic involution and immune reconstitution. Trends Immunol 30:366–373
6. Douek DC, McFarland RD, Keiser PH, Gage EA, Massey JM, Haynes BF, Polis MA, Haase AT, Feinberg MB, Sullivan JL, Jamieson BD, Zack JA, Picker LJ, Koup RA (1998) Changes in thymic function with age and during the treatment of HIV infection. Nature 396:690–695
7. Kong F, Chen CH, Cooper MD (1998) Thymic function can be accurately monitored by the level of recent T cell emigrants in the circulation. Immunity 8:97–104
8. Douek DC, Vescio RA, Betts MR, Brenchley JM, Hill BJ, Zhang L, Berenson JR, Collins RH, Koup RA (2000) Assessment of thymic output in adults after haematopoietic stem-cell transplantation and prediction of T-cell reconstitution. Lancet 355:1875–1881
9. Davis CC, Marti LC, Sempowski GD, Jeyaraj DA, Szabolcs P (2010) Interleukin-7 permits Th1/Tc1 maturation and promotes ex vivo expansion of cord blood T cells: a critical step toward adoptive immunotherapy after cord blood transplantation. Cancer Res 70:5249–5258
10. Page KM, Mendizabal AM, Prasad VK, Martin PL, Parikh S, Wood S, Sempowski GD, Szabolcs P, Kurtzberg J (2008) Posttransplant autoimmune hemolytic anemia and other autoimmune cytopenias are increased in very young infants undergoing unrelated donor umbilical cord blood transplantation. Biol Blood Marrow Transplant 14:1108–1117
11. Liu C, Chen BJ, Deoliveira D, Sempowski GD, Chao NJ, Storms RW (2010) Progenitor cell dose determines the pace and completeness of engraftment in a xenograft model for cord blood transplantation. Blood 116:5518–5527
12. Sarzotti-Kelsoe M, Win CM, Parrott RE, Cooney M, Moser BK, Roberts JL, Sempowski GD, Buckley RH (2009) Thymic output, T-cell diversity, and T-cell function in long-term human SCID chimeras. Blood 114:1445–1453
13. Sempowski GD, Gooding ME, Liao HX, Le PT, Haynes BF (2002) T cell receptor excision circle assessment of thymopoiesis in aging mice. Mol Immunol 38:841–848
14. Hockett RD Jr, Nunez G, Korsmeyer SJ (1989) Evolutionary comparison of murine and human delta T-cell receptor deleting elements. The New Biologist 1:266–274
15. de Villartay JP, Hockett RD, Coran D, Korsmeyer SJ, Cohen DI (1988) Deletion of the human T-cell receptor delta-gene by a site-specific recombination. Nature 335:170–174
16. Toda M, Fujimoto S, Iwasato T, Takeshita S, Tezuka K, Ohbayashi T, Yamagishi H (1988) Structure of extrachromosomal circular DNAs excised from T-cell antigen receptor alpha and delta-chain loci. J Mol Biol 202:219–231
17. Takeshita S, Toda M, Yamagishi H (1989) Excision products of the T cell receptor gene support a progressive rearrangement model of the alpha/delta locus. EMBO J 8:3261–3270
18. Elliott JF, Rock EP, Patten PA, Davis MM, Chien YH (1988) The adult T-cell receptor delta-chain is diverse and distinct from that of fetal thymocytes. Nature 331:627–631
19. Min B, McHugh R, Sempowski GD, Mackall C, Foucras G, Paul WE (2003) Neonates support lymphopenia-induced proliferation. Immunity 18:131–140
20. Chen BJ, Cui X, Sempowski GD, Gooding ME, Liu C, Haynes BF, Chao NJ (2002) A comparison of murine T-cell-depleted adult bone marrow and full-term fetal blood cells in hematopoietic engraftment and immune reconstitution. Blood 99:364–371
21. Billard MJ, Gruver AL, Sempowski GD (2011) Acute endotoxin-induced thymic atrophy is characterized by intrathymic inflammatory and wound healing responses. PloS One 6:e17940
22. Shi X, Zhang P, Sempowski GD, Shellito JE (2011) Thymopoietic and bone marrow response to murine Pneumocystis pneumonia. Infect Immun 79:2031–2042
23. Chen BJ, Deoliveira D, Spasojevic I, Sempowski GD, Jiang C, Owzar K, Wang X, Gesty-Palmer D, Cline JM, Bourland JD, Dugan G, Meadows SK, Daher P, Muramoto G, Chute JP, Chao NJ (2010) Growth hormone mitigates against lethal irradiation and enhances hematologic and immune recovery in mice and nonhuman primates. PloS One 5:e11056
24. Gruver AL, Ventevogel MS, Sempowski GD (2009) Leptin receptor is expressed in thymus medulla and leptin protects against thymic remodeling during endotoxemia-induced thymus involution. J Endocrinol 203:75–85
25. Sodora DL, Douek DC, Silvestri G, Montgomery L, Rosenzweig M, Igarashi T, Bernacky B, Johnson RP, Feinberg MB, Martin MA, Koup RA (2000) Quantification of thymic function by measuring T cell receptor excision circles within peripheral blood and lymphoid tissues in monkeys. Eur J Immunol 30:1145–1153
26. Bona R, Macchia I, Baroncelli S, Negri DR, Leone P, Pavone-Cossut MR, Catone S,

Buffa V, Ciccozzi M, Heeney J, Fagrouch Z, Titti F, Cara A (2007) T cell receptor excision circles (TRECs) analysis during acute intrarectal infection of cynomolgus monkeys with pathogenic chimeric simian human immunodeficiency virus. Virus Res 126:86–95

27. Aspinall R, Pido-Lopez J, Imami N, Henson SM, Ngom PT, Morre M, Niphuis H, Remarque E, Rosenwirth B, Heeney JL (2007) Old rhesus macaques treated with interleukin-7 show increased TREC levels and respond well to influenza vaccination. Rejuvenation Res 10:5–17
28. Muthukumar A, Zhou D, Paiardini M, Barry AP, Cole KS, McClure HM, Staprans SI, Silvestri G, Sodora DL (2005) Timely triggering of homeostatic mechanisms involved in the regulation of T-cell levels in SIVsm-infected sooty mangabeys. Blood 106:3839–3845
29. Sodora DL, Milush JM, Ware F, Wozniakowski A, Montgomery L, McClure HM, Lackner AA, Marthas M, Hirsch V, Johnson RP, Douek DC, Koup RA (2002) Decreased levels of recent thymic emigrants in peripheral blood of simian immunodeficiency virus-infected macaques correlate with alterations within the thymus. J Virol 76:9981–9990

Chapter 13

Isolation of RNA and the Synthesis and Amplification of cDNA from Antigen-Specific T Cells for Genome-Wide Expression Analysis

R. Anthony Barnitz, Sabrina Imam, Kathleen Yates, and W. Nicholas Haining

Abstract

Genome-wide gene expression analysis has become a very powerful routine tool for the study of distinct differentiation states. However, the examination of total populations of cells that contain high levels of heterogeneity, such as the total CD8+ T cell population during an immune response, is limited because that complexity hampers accurate interpretation. The gene expression signatures from populations represent the average of all cells within the populations, which will smooth out large expression changes within small subpopulations and virtually eliminate any small changes. However, small expression changes within a minor subpopulation, such as antigen-specific CD8+ T cells responding to an infection, can have relevant biological consequences. Although very limited amounts of RNA can be isolated from small subpopulations of cells, there are now methods to synthesize and amplify cDNA from this limited RNA in sufficient quantities needed for microarray analysis. Here, we describe a complete protocol to extract RNA from small numbers of cells, synthesize cDNA from that RNA, and amplify that cDNA in an unbiased method. This protocol is a useful tool for the study of genome-wide expression signatures from many of the subpopulations that are numerically small but important in immune responses and homeostasis.

Key words: Genome-wide gene expression analysis, Antigen-specific T cells, RNA extraction, cDNA synthesis, cDNA amplification, Single primer isothermal amplification (SPIA)

1. Introduction

Genome-wide expression profiling has now become a routine tool to identify patterns of genes characteristic of different differentiation states. Large-scale maps of the differentiation stages within the human hematopoietic system have recently been assembled in order to identify candidate regulatory mechanisms that specify

Andrew L. Snow and Michael J. Lenardo (eds.), *Immune Homeostasis: Methods and Protocols*, Methods in Molecular Biology, vol. 979, DOI 10.1007/978-1-62703-290-2_13, © Springer Science+Business Media New York 2013

lineage choices (1). These maps are also being assembled in the mouse (2). However, the analysis of differentiation states in increasingly granular populations comes at a cost of smaller numbers of cells contained in each subpopulation. There is therefore a need for robust methods with which to extract, purify, and amplify the small amounts of RNA present in tens or hundreds of cells.

In the immune system, microarray analysis of populations of cells has proven to be a powerful discovery tool in immunology (3). However, the use of total populations of cells for microarray analysis is limited in two critical ways. First, the analysis of bulk populations may entrain cells with a similar surface-marker phenotype but unrelated biology. Second, the expression levels of each gene measured in a population of cells represents the average of all cells in the sample. As such, analysis of cell ensembles cannot distinguish between homogenous expression of a gene in many cells or extreme expression levels in a few. Heterogeneity is the hallmark of the adaptive immune system (4). Marked heterogeneity has been described for both $CD4^+$ and $CD8^+$ T cells (5, 6). Therefore, because only the cells specific for disease antigens become activated in an immune response, the study of the total population of $CD8^+$ T cells is likely to only capture large changes in expression within the small subpopulation of responding cells. This limits the overall analysis from a genome-wide approach to a study of fewer genes with large differential expression. This becomes particularly important when considering the fact that relatively small and gradual changes in critical transcription factors may be associated with marked changes in cell fate (1). Biologically relevant changes in gene expression can therefore be masked by analysis of bulk populations of cells rather than discrete subpopulations.

The ability to measure complex gene expression signatures in small numbers of antigen-specific cells is therefore critical to understanding mechanisms dictating cell fate. We have recently used the protocol outlined in this chapter to examine $CD8^+$ T cell exhaustion during HIV-1 infection (7). By specifically studying HIV-specific $CD8^+$ T cells from progressors versus elite controllers, we were able to identify the upregulation of BATF, a transcription factor in the AP-1 family, in exhausted cells. Another group has used a similar approach to study $CD8^+$ T cell exhaustion in the context of cancer (8). However, although the melanoma-specific cells from this study exhibited similar gene expression signatures compared to exhausted LCMV-specific $CD8^+$ T cells (9), exhausted tumor infiltrating cells did not show an upregulation in BATF, suggesting that there may be multiple mechanisms of T cell exhaustion depending on the context of the disease. Here, we provide a detailed procedure to isolate RNA from small numbers of cells, and then synthesize and amplify cDNA to quantities suitable for microarray analysis.

2. Materials

2.1. Work Area Cleaning and Decontamination

1. DNA Off (MP Biomedicals).
2. RNase ZAP (Ambion).
3. Ethanol (EtOH).

2.2. RNA Extraction and Purification

1. Siliconized microcentrifuge tubes (Thermo Fisher Scientific).
2. TRIzol (Invitrogen).
3. Phase-Lock Heavy Gel Tubes (5 Prime).
4. Chloroform.
5. Agencourt RNAdvance Tissue Isolation kit (Beckman Coulter) (see Note 1).
6. Isopropanol.
7. Abgene 1.2 ml Square Well Storage plate (Abgene).
8. Agencourt SPRIPlate 96R—Ring Super Magnet Plate (Beckman Coulter).
9. Ethanol.
10. DNase/RNase-free water.
11. 10× DNase I buffer (Applied Biosystems).
12. DNase I (Applied Biosystems).
13. Abgene PCR Sealer (Thermo Fisher Scientific).

2.3. cDNA Synthesis and Amplification

1. Ovation Pico WTA System V2 (12 rx) (NuGEN, Inc).
2. 0.2 ml PCR plate (see Note 2).
3. Agencourt SPRIPlate 96R—Ring Super Magnet Plate (Beckman Coulter).
4. Abgene PCR Sealer (Thermo Fisher Scientific).
5. Zymo Research DNA Clean & Concentrator-25 (Zymo Research).
6. Ethanol.

3. Methods

3.1. RNA Extraction

1. Sort HLA-tetramer^{+} antigen-specific T cells into siliconized microcentrifuge tubes using an appropriate staining and sorting procedure (7). Pellet the cells and aspirate the supernatant. Because the cell pellet may not be visible, be conservative when aspirating. A small amount of supernatant left with the pellet will not adversely affect the extraction of the RNA. Resuspend

the cells in 1 ml of TRIzol by pipetting up and down several times (see Note 3). Freeze the TRIzol mixture at −80°C until ready to extract the RNA.

2. The work area and pipettes should be cleaned with DNA Off, RNase ZAP, and 70% EtOH prior to starting work. Always clean gloves with alcohol, and change them frequently, especially if you touch uncleaned, potentially RNase contaminated surfaces (see Note 4).
3. Obtain samples frozen in TRIzol from −80°C, and arrange them randomly (to minimize any batch effects during downstream applications).
4. Let samples stand 10–15 min at room temperature post-thawing before proceeding. Ensure that the samples are well-homogenized.
5. Spin phase-lock heavy gel tubes in a microcentrifuge for 30 s at ~13,000 × *g* to bring the gel to the bottom of the tubes.
6. Transfer the contents of the siliconized tubes, containing cells in TRIzol (approximately 1 ml), to the phase lock tubes, being careful not to disturb the gel. Add 200 μl of chloroform (per 1 ml TRIzol) to each tube, and mix by inversion for 15 s. Incubate at room temperature for 3 min.
7. Spin phase-lock tubes containing samples at ~14,000 × *g* for 20 min at 4°C (during this spin, prepare steps 1–4 in Subheading 3.2). The top (clear) aqueous phase should be above the gel plug, and the pink organic phase should be underneath it (see Note 5).

3.2. RNA Purification

We use the Agencourt RNAdvance Tissue Isolation kit (see Note 1). During the 20 min spin:

1. Take the Bind Buffer (from the kit) out of the refrigerator, and allow it to warm to room temperature.
2. For the first time using the kit, be sure to add the appropriate volume of 100% isopropanol to the wash buffer (see kit protocol for specific volume to add). Mix thoroughly. This can be stored at room temperature for 6 months.
3. Prepare the Bind Solution (prepare fresh for each experiment and discard unused portions). For each sample mix 80 μl of Bind Buffer and 450 μl of Isopropanol.
4. Add 100 μl Lysis Buffer (from kit) to wells of an Abgene 1.2 ml Square Well Storage plate once the spin (Subheading 3.1, step 7) has been completed.
5. Transfer only 500 μl of the aqueous phase from the TRIzol/chloroform extraction of each sample to a well already containing

Lysis Buffer. Pipet to mix. The resulting solution should appear opaque white.

6. Make sure the Bind Solution (prepared in step 3) is resuspended (the beads can settle very quickly; pipet to mix). Add 530 μl of Bind Solution to each well and slowly pipet 10 times, ensuring that the solution is well mixed. Incubate at room temperature for 5 min (see Note 6).
7. Place the plate on top of the Agencourt SPRIPlate 96R—Ring Super Magnet Plate for 10 min. Wait for the solution to clear and the beads to appear as a ring on the bottom of the well (see Note 7).
8. With the plate still on the magnet, fully remove supernatant from each well and discard (see Note 8).
9. Remove the plate from the magnet. Wash the beads by adding 800 μl of the Wash Buffer (from the kit; containing isopropanol). Slowly pipet 10 times, ensuring the solution is well mixed. Try to avoid bubbles while mixing.
10. Place the plate on the magnet for 5 min. After this incubation, fully remove the supernatant from each well, checking for beads in the tip used to aspirate the supernatant, and discard.
11. Remove the plate from the magnet. Wash by adding 800 μl of freshly prepared 80% EtOH (using DNase/RNase free water). Gently pipet 4 times to mix.
12. Place the plate on the magnet for 5 min. After this incubation, fully remove the supernatant from each well, check for beads, and discard. Let the plate sit on the magnet for approximately 5 min for residual EtOH evaporation before removing the plate from the magnet (see Note 9).
13. Prepare DNase I Solution using the 10× DNase I buffer and DNase I. For each sample mix 10 μl of DNase I, 10 μl 10× DNase I buffer, and 80 μl of DNase/RNase-free water.
14. Remove the plate from the magnet. Add 100 μl of the DNase I solution to each well and incubate at room temperature for 1 min without mixing to hydrate the beads. After this incubation, gently pipet 5 times to resuspend the beads in the DNase I solution. Seal the plate using a PCR Sealer, and incubate the plate at 37°C for 15 min.
15. After the incubation, do not remove the DNase I solution. Add 550 μl Wash Buffer and pipet 5 times to mix. Incubate at room temperature for 4 min.
16. Place the plate on the magnet for 7 min. Wait for the solution to clear before proceeding.

17. Carefully aspirate the supernatant and wash by adding 600 μl of 80% EtOH. Do not pipet to mix. This step is performed on the magnet.
18. Incubate for 2 min to allow the beads to resettle. Remove EtOH from each well, check for beads, and discard.
19. Repeat steps 17-18 twice for a total of three EtOH washes.
20. Remove EtOH from the final wash completely and allow the beads to dry for 15 min at room temperature. The beads do not need to be completely dry, but all traces of liquid should be gone.
21. Remove the plate from the magnet and elute by adding 16 μl DNase/RNase-free water (or desired volume for elution). Pipet 10 times to mix and incubate at room temperature for 2 min (see Note 10).
22. Return the plate to the magnet for 2 min and carefully transfer the eluted RNA away from the beads and into nuclease-free microcentrifuge tubes for storage at −80°C (see Note 11).

3.3. First Strand cDNA Synthesis

We use the Ovation Pico WTA System V2 from NuGEN, Inc. to synthesize and amplify cDNA from the small amounts of RNA isolated from small numbers of antigen-specific T cells. The work area and pipettes should be cleaned with DNA Off, RNase ZAP, and 70% EtOH prior to starting work. Always clean gloves with alcohol, and change them frequently, especially if you touch potentially contaminated surfaces. Please note that NuGEN recommends that the work area be cleaned 1–2 days prior to using the kit to allow for any aerosolized particles of DNase/RNase to settle (see Note 12).

1. Obtain the First Strand Primer Mix (Tube A1), First Strand Buffer Mix (Tube A2), First Strand Enzyme Mix (A3), and the RNase-free water (D1) from −20°C, and thaw the solutions.
2. Flick tube A3 to mix the contents. Then spin down the tube at ~14,000 × *g* for 2 s and place on ice.
3. Thaw the additional reagents (Tubes A1, A2, and D1) at room temperature. Mix the tubes by vortexing for 2 s. Then, spin down the tubes and place on ice. Leave the water (D1) at room temperature.
4. Add 2 μl of A1 to each assigned well of a 0.2 ml PCR plate on ice (make sure to dispense at the bottom of the well).
5. Add 5 μl of total RNA sample to the primer in the appropriate well. Pipet to mix (see Note 13).
6. Seal and spin the plate briefly. Return the plate to ice.
7. Put the plate in the prewarmed thermocycler, and incubate the plate at 65°C for 2 min, followed by a hold at 4°C. (see Note 14).

8. While the plate is incubating, prepare an enzyme master mix (mix by pipetting, spin down and keep on ice). For each sample (well in the plate), mix 2.5 μl of the First Strand Buffer Mix (A2) and 0.5 μl of the First Strand Enzyme Mix (A3).
9. After the plate has finished the incubation, remove it from the thermocycler, spin it down, and place it on ice.
10. Add 3 μl of the Master Mix from step 8 to each well. Mix by pipetting 5 times, seal the plate, and spin down the plate.
11. Put the plate on ice until transfer to the thermocycler. Incubate the plate in the thermocycler using the following stages: 4°C for 2 min, 25°C for 30 min, 42°C for 15 min, 70°C for 15 min, and hold at 4°C. (Approx. halfway through, see steps 1–4 of the Second Strand cDNA Synthesis Protocol: Subheading 3.4).
12. Once the incubations are finished, spin down the plate, and put it on ice.

3.4. Second Strand cDNA Synthesis

1. Remove the Agencourt RNAClean XP purification beads (from the NuGEN kit) from 4°C and place them on the benchtop to reach room temperature (~15 min).
2. Obtain the Second Strand Buffer Mix (B1) and Second Strand Enzyme Mix (B2) from −20°C.
3. Flick tube B2 to mix, spin down the contents for 2 s, and place it on ice.
4. Thaw tube B1 at room temperature, vortex for 2 s, and spin down the contents. Put the tube on ice.
5. Prepare an enzyme Master Mix (mix by pipetting, spin down and keep on ice). For each sample (well), mix 9.7 μl of the Second Strand Buffer Mix (B1) and 0.3 μl of Second Strand Enzyme Mix (B2).
6. Add 10 μl of the Second Strand Master Mix to each well. Mix each well by pipetting 5 times. Seal the plate, spin it down, and place it on ice.
7. Transfer the plate to the thermocycler, and incubate the plate using the following stages: 4°C for 1 min, 25°C for 10 min, 50°C for 30 min, 80°C for 20 min, and hold at 4°C (see Note 14).
8. Once the incubations are finished, spin down the plate to collect condensation, and put it on ice. Continue immediately to Purification of cDNA (Subheading 3.5).

3.5. Purification of cDNA

1. Ensure that the Agencourt RNAClean XP beads have reached room temperature. Shake well (invert and tap) to mix and ensure an even resuspension of the beads.

2. Prepare a 70% EtOH wash solution. It is critical that the solution be made fresh on the day of the experiment and that it is well mixed.
3. Add suspension to each reaction well and mix by pipetting up and down 10 times. Incubate at room temperature for 10 min.
4. Place the plate on top of the Agencourt magnetic plate and let stand at least 5 min to completely clear the solution of beads.
5. Carefully remove 45 μl of the supernatant and discard. When removing the supernatant, insert the pipet tip to the bottom of the well, being careful not to scrape the sides of the well (see Note 15).
6. With the plate still on the magnet, add 200 μl of freshly prepared 70% EtOH to each well. Allow the plate to stand for 30 s before removing. The beads will not disperse; they should remain in a ring on the walls of the well. Significant loss of beads at this stage will impact cDNA yields, so ensure beads are not removed with the binding buffer or the washes.
7. Repeat the 70% EtOH wash 2 more times. When removing the final wash, be careful to remove as much EtOH as possible, so as to reduce the drying time. Use at least two additional removal steps to allow EtOH to flow down to the bottom of the well for removal.
8. Air-dry the beads on the magnet for a minimum of 15–20 min. Inspect each well/tube carefully to ensure that all the EtOH has evaporated. When the beads change color from dark brown to light brown, the residual EtOH has evaporated. All residual EtOH must be removed prior to the single primer isothermal amplification (SPIA) procedure. Continue immediately to the Amplification of the cDNA (Subheading 3.6).

3.6. Amplification of the cDNA

1. Obtain the SPIA Primer Mix (C1), SPIA buffer mix (C2) and SPIA Enzyme mix (C3) stored at −20°C.
2. Thaw reagents C1 and C2 at room temperature, mix by vortexing briefly, and spin down. Put the tubes on ice.
3. Thaw C3 on ice and mix contents by inverting gently 5 times. Ensure the enzyme is well mixed without creating bubbles (enzymes denature at air-water interfaces), and then spin the tube down before placing it on ice.
4. Make a SPIA Master Mix, and put it on ice. Please note that the order of addition is very important (add the reagents in the order they are listed). For each sample (well), mix 50 μl of SPIA Buffer Mix (C2), 25 μl of SPIA Primer Mix (C1), and 25 μl of SPIA Enzyme mix (C3).

5. Add 100 µl of the SPIA master mix to each well. Use a pipet to mix well by pipetting up and down (8–10 times), and carefully wash the sides of the well. Attempt to get the majority of beads in suspension (may not be uniform, but that is fine because the SPIA master mix will elute the cDNA from the beads).
6. Seal the plate, and put it on ice. Transfer the plate to the thermocycler, and incubate the plate using the following stages: 4°C for 1 min, 47°C for 75 min, 95°C for 5 min, and hold at 4°C (see Note 14).
7. Remove the plate from the thermocycler, spin it down, and put it on ice. Do NOT open the plate in the pre-amplification workspace. All remaining steps should be performed in the post-amplification workspace.
8. Transfer the plate to the magnetic plate and let stand for 5 min to completely clear the solution of beads.
9. Carefully remove the cleared supernatant containing the eluted cDNA and transfer to a fresh tube/plate. The beads may be discarded.
10. At this stage the cDNA can be purified or stored at −20°C.

3.7. Purification of the Amplified cDNA

For this procedure we use the DNA Clean and Concentrator-25 from Zymo Research (Cat# 4005 or 4006). Ordering the correct product is important; Cat# 4033 or 4034 will result in low yields of cDNA.

1. Obtain and label a new nuclease free 1.5 ml microcentrifuge tube for each sample.
2. Add 320 µl DNA Binding Buffer (from kit) to each tube, and then add 160 µl SPIA cDNA product. Vortex briefly and spin down the tubes.
3. For each sample, place a Zymo-Spin II column in a collection tube.
4. Load the entire volume of sample from step 2 (480 µl) onto the Zymo-Spin II column.
5. Centrifuge at ~13,000 × *g* rpm for 10 s (see Note 16). Discard the flow-through. Replace the column in the same collection tube.
6. Wash sample with 200 µl of 80% EtOH (freshly prepared, at room temperature). Do not use the wash buffer provided in the kit.
7. Centrifuge at ~13,000 × *g* for 10 s. Discard the flow-through.
8. Add 200 µl of room temp 80% EtOH.
9. Centrifuge the column for 30 s at ~13,000 × *g*. Discard the flow-through.
10. Blot the tip of the column on filter paper to remove any residual EtOH. Place the column in a new labeled nuclease-free 1.5 ml microcentrifuge tube.

11. Add 30 μl of room temperature nuclease-free H_2O (D1 from the NuGEN kit) to the center of each column. Do not use cold water.
12. Centrifuge the column at ~13,000 × *g* for 30 s.
13. Discard the column and collect the sample.
14. Mix the sample by vortexing, then spin briefly. Store the sample at −20°C.

3.8. Quality Control Testing of the Amplified cDNA

1. Measure the concentration and the A260/A280 ratio of the cDNA using a spectrophotometer (e.g., NanoDrop).
2. Submit an aliquot of representative samples for analysis on a BioAnalyzer RNA Nano 6000 chip (see Note 17).
3. Amplify a control gene by RT-PCR (note: use ~20 ng of input cDNA).

4. Notes

1. An important consideration in the refinement of RNA extraction protocols was the low cell numbers. Since T cells typically contain less than half a picogram of RNA per cell, when only a few thousand (and in rarer cases, a few hundred) tetramer$^+$ cells are isolated, keeping the eluted RNA as concentrated as possible is a priority, in order to maximize the efficiency of downstream applications. We considered QIAGEN RNeasy columns, but found that when elution volumes of 10–14 μl were used, we typically lost at least 2 μl to "dead" volume in the columns, thus adversely affecting our total yields. We also considered a simple TRIzol/chloroform extraction with a carrier such as glycogen; however, this was limited by lower purity of RNA for downstream applications. Ultimately, this adapted Agencourt RNA extraction protocol was the most suitable for our purposes; it included a DNase digestion step (genomic DNA contamination is a serious complication in the NuGEN amplification steps), a scalable elution volume with minimal loss, and yielded RNA at the greatest purity and yield in our side-by-side comparisons. Additionally, when extracting RNA from large numbers of samples, the 96-well plate format allowed us to process greater numbers of samples in fewer batches.
2. It is advantageous to use a PCR plate over tube strips because a plate has better stability when sitting on top of the magnetic plate. When choosing a PCR plate to use, look for a plate that fits well into your thermocycler and that fits well into the magnetic plate. The purification of cDNA steps prior to amplification

require that magnetic beads binding the cDNA be washed as the PCR plate sits on top of the magnetic plate. At this time, the beads form a ring around the well of the PCR plate. It is preferable to select a PCR plate in which this ring of beads will appear at the middle of the well, as opposed to the base of the well. Having the beads form a ring at the middle of the well will decrease the likelihood of losing beads during aspiration in the wash steps.

3. Use siliconized microcentrifuge tubes when collecting small numbers of cells for subsequent RNA isolation. After collection, pellet the cells and resuspend them in 1 ml of TRIzol, regardless of the cell number. This will ensure that the volume of the aqueous phase (and concentration of all other reagents) remains constant for all samples.
4. Clean work area at least 1 h (best is overnight) prior to beginning the protocol to prevent aerosolized cleaning agents from adversely affecting RNA yields.
5. The aqueous phase should be as close to clear as possible. If it appears slightly pink, the phase separation likely did not occur as efficiently as desired. However, this should not affect the purification, as there are many wash steps in the protocol that will remove any residual phenol that could harm downstream applications. If the aqueous phase appears very pink, or if the gel plug does not form an adequate barrier between the phases, carefully remove as much of the TRIzol/chloroform mixture from the existing phase lock tube as possible, and place this mixture in a new phase-lock tube that has already been spun-down such that the gel is at the bottom of the tube. Repeat steps 4 and 5, but do not add additional chloroform.
6. The total volume in each well at this stage is 1,130 μl. Use caution to avoid spillover of contents between wells. In addition, the beads tend to aggregate at this stage. Mix until no aggregation is visible. While mixing the aqueous phase and the bind solution, pipet from the bottom of the well to the top to allow for a homogenous solution.
7. If there is insufficient mixing during step 11, a distinct separation between the aqueous phase and the bead/bind solution will be evident (the solution will appear brown as opposed to clear, and the beads will not settle to the bottom of the well). In this case, remove the plate from the magnet, mix for an additional 2 min, and then replace the plate on the magnet for 10 min.
8. Avoid disturbing the separated magnetic beads by aiming the pipet tip for the center of the well bottom. Examine the contents of the tips before discarding. If beads are drawn out, replace the supernatant in the well and give the beads 2 min to settle on the

magnet, then attempt to remove the supernatant again. If necessary, leave a few microliters of supernatant behind.

9. Remove as much EtOH as possible. Residual EtOH will inhibit the DNase I digestion.
10. The elution volume suggested here was selected in order to allow for three potential rounds of reverse-transcription, using the NuGEN Ovation Pico WTA System V2. This volume can be adjusted based on the user's downstream uses of the RNA.
11. RNA quality can be assessed using a NanoDrop (if the starting cell number was greater than 50,000 cells, i.e. an RNA concentration greater than 2 ng/microliter), or a BioAnalyzer RNA Nano or Pico chip. For very small cell numbers (less than 5,000), it is advised to proceed straight to reverse-transcription without attempting to quantify the RNA, as it will likely be below the limits of detection.
12. To avoid carrying-over and amplifying non-specific products, NuGEN recommends that one physically separate the pre-amplification and post-amplification steps. If possible, designate two opposite areas of the lab for pre- and post-amplification steps. Use separate instruments, or thoroughly clean all pipettes and magnets, used between the two stages of the protocol.
13. NuGEN recommends starting with 500 pg to 50 ng of RNA. If starting with a large number of cells, one may wish to quantify the RNA after extraction and to normalize the RNA input across samples for reverse-transcription. However, if one begins with less than 20,000 cells, the RNA quantity may be difficult to assess accurately. As a result, it is recommended here that one use 5 μl of RNA input for all samples. Later, cDNA will be normalized across samples for fragmentation, labeling, and hybridization to microarrays.
14. For all programs using the thermocycler, allow the plate warmer to cool before putting the plate into the thermocycler. When the plate warmer has reached the correct temperature, pause the program, place the plate in the thermocycler, and resume.
15. Aspirate at a slow, steady rate to avoid accidental removal of beads. Prior to discarding buffer, check the tip(s) to make sure that you have not removed any beads. If you observe any beads on/in a tip, replace the volume in the well and let sit for another 5 min in the presence to the magnet and reattempt.
16. Begin the timer only when the microcentrifuge has reached the desired speed.
17. In interpreting BioAnalyzer results for NuGEN-amplified cDNA, use the Eukaryotic total RNA Nano program and follow the manufacturer's instructions. Denaturation of the SPIA cDNA sample (as described in the Agilent protocol) is required

for optimum resolution. The shape of the size distribution trace is dependent on the input RNA integrity and source, and it may vary significantly. Contact Nugen technical support with questions. A peak or odd tail at higher molecular weights could indicate genomic DNA contamination.

Acknowledgments

This work was supported by the National Institutes of Health (NIAID R01AI091493 to W.N.H.).

References

1. Novershtern N et al (2011) Densely interconnected transcriptional circuits control cell states in human hematopoiesis. Cell 144:296–309
2. Heng TS, Painter MW (2008) The Immunological Genome Project: networks of gene expression in immune cells. Nat Immunol 9:1091–1094
3. Haining WN, Wherry EJ (2010) Integrating genomic signatures for immunologic discovery. Immunity 32:152–161
4. Kaech SM, Wherry EJ (2007) Heterogeneity and cell-fate decisions in effector and memory CD8+ T cell differentiation during viral infection. Immunity 27:393–405
5. Lefrancois L, Marzo AL (2006) The descent of memory T-cell subsets. Nat Rev Immunol 6:618–623
6. Roman E et al (2002) CD4 effector T cell subsets in the response to influenza: heterogeneity, migration, and function. J Exp Med 196: 957–968
7. Quigley M et al (2010) Transcriptional analysis of HIV-specific CD8+ T cells shows that PD-1 inhibits T cell function by upregulating BATF. Nat Med 16:1147–1151
8. Baitsch L et al (2011) Exhaustion of tumor-specific CD8 T cells in metastases from melanoma patients. J Clin Invest 121:2350–2360
9. Wherry EJ et al (2007) Molecular signature of CD8+ T cell exhaustion during chronic viral infection. Immunity 27:670–684

Chapter 14

Designs for Massively Parallel Sequencing Approaches to Identify Causal Mutations in Human Immune Disorders

Yu Zhang and Helen C. Su

Abstract

Massively parallel sequencing technologies provide new opportunities to discover causal variants and narrow down candidate genes responsible for human Mendelian disorders. Such information can in turn provide new insights into understanding the basic science behind, as well as improving diagnosis and treatment for, these disorders. In this chapter, we review experimental design and data analysis for sequencing studies of human immune disorders. We discuss optimal experimental designs for sample selection and sequencing approaches, as well as key aspects of data analysis such as filtering and prioritization of identified variants.

Key words: Mendelian diseases, Genetic Inheritance Patterns, Exome sequencing, Whole genome sequencing, Causal variants, Filtering and prioritization of variants

1. Introduction

Recent advances in massively parallel DNA sequencing technologies, especially exome sequencing, have provided novel approaches for discovering new causal variants and narrowing down candidate genes for inherited human diseases. Previously, identifying causative genetic loci for Mendelian disorders was generally performed by linkage mapping or association studies followed by positional cloning or candidate gene analyses. However, this strategy is limited to cases with large pedigrees that contain multiple affected individuals or foreknowledge about likely genes/pathways that are likely to be involved. By contrast, the newer sequencing technologies are particularly suitable for unbiased investigation of rare disorders, which usually lack sufficient statistical power for linkage studies. These technologies enable rapid sequencing of all genes or the whole genome of any individual, thereby revealing DNA variations at an unprecedented resolution and scale. The resulting profile of

Andrew L. Snow and Michael J. Lenardo (eds.), *Immune Homeostasis: Methods and Protocols*, Methods in Molecular Biology, vol. 979, DOI 10.1007/978-1-62703-290-2_14, © Springer Science+Business Media New York 2013

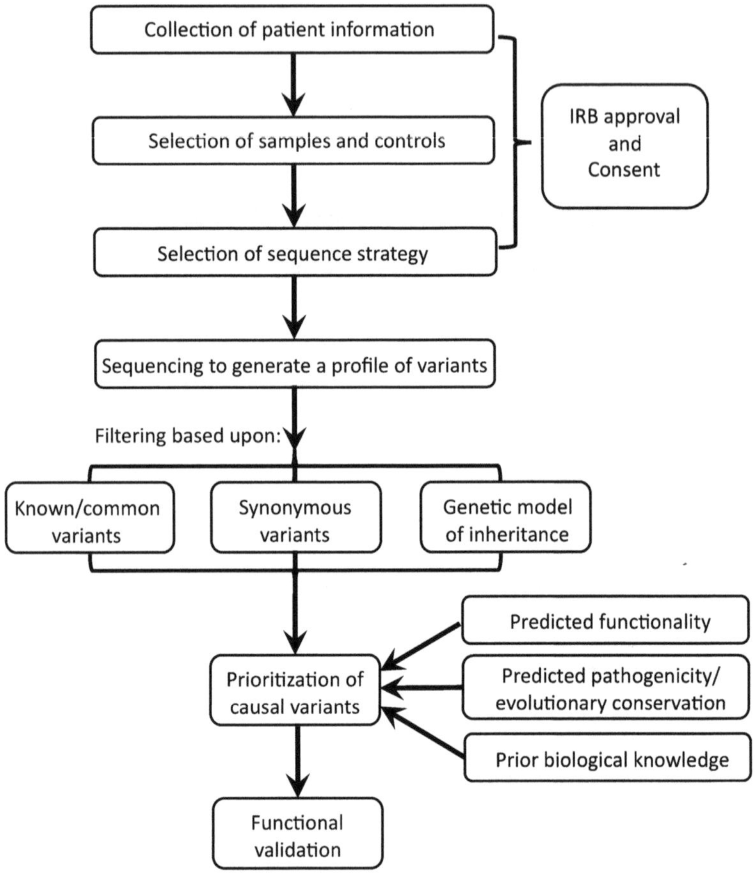

Fig. 1. Workflow for using massively parallel sequencing to identify causal disease gene discoveries.

variants from people who have a particular disease offers the potential to identify causal genetic disease genes, particularly for rare Mendelian immune disorders. Even in non-Mendelian or complex immune diseases, the genetic makeup of the affected individuals still contributes more or less to disease manifestations. However, the application of massively parallel DNA sequencing technologies to the study of non-Mendelian or complex immune diseases is still in its infancy and will not be discussed further.

In this chapter, we review the conceptual framework of such an approach as applied to Mendelian diseases of the immune system. The workflow we discuss is summarized in Fig. 1. Rather than focusing on the sequencing techniques themselves, which are rapidly changing, we will focus on the practical aspects of experimental design and efficient analysis of the data sets. Our goal is to provide useful information and guidance for those who have limited experience, but plan to use this technique in their own experiments.

2. Models of Genetic Inheritance

Medical geneticists traditionally gather information to construct family pedigrees for help in diagnosis and genetic counseling. This information is also important in setting the groundwork for research studies in which linkage and positional cloning are used to pinpoint the genetic locus associated with disease in large families. Although newer sequencing strategies are powerful and decrease the need to collect large families or numbers of patients, knowing the genetic model of inheritance is still important for the data analysis and interpretation of the variant profiles found by sequencing. Thus, efforts should be directed towards collecting complete clinical records and a detailed family history of the patients. This data can be reviewed in order to tease apart genetic and environmental factors that contribute to disease pathogenesis, as well as to identify the actual or likely inheritance patterns of the disease. Such information will help investigators to select the best experimental and computational strategies for sequencing and identification of causative genetic variants.

2.1. Autosomal Recessive Inheritance

The typical inheritance pattern for an autosomal recessive disease is that affected children, of either sex, make up ~25% of the offspring of unaffected healthy parents. Whereas both parents are expected to have one defective copy and one normal copy, a patient inherits two defective copies of the same autosomal gene, one from each parent. The affected patient carries either a homozygous mutation or two different deleterious mutations in the same gene (compound heterozygous mutations). For a patient from a family with known consanguinity or from a genetically isolated population, a homozygous mutation usually makes up the causal variant since the parents are related and share identical copies of some genes. By contrast, for a patient having unrelated parents, the patient is more likely to carry compound heterozygous mutations. The number of homozygous (or compound heterozygous) mutations is significantly lower than the number of heterozygous mutations that can be found in any individual's genome. These homozygous or compound heterozygous variants are usually premature termination or splicing acceptor and donor site mutations, leading to loss of function when both copies are defective. In addition to the limited number of potential causal variants, the ease in identifying and experimentally validating loss-of-function mutations creates a natural bias towards detecting mutations responsible for recessive disease. So far, most of the disease-causing genes successfully discovered by sequencing approaches so far belong to this category. For example, Bolze et al. recently identified a homozygous missense mutation in FADD (Fas-associated death domain protein), in patients having an autosomal recessive disease that resembles the autoimmune lymphoproliferative syndrome (ALPS) (1).

2.2. Autosomal Dominant Inheritance

The typical inheritance pattern for an autosomal dominant disease is that of multiple affected individuals, of either sex, making up ~50% of the offspring. Although it is particularly easy to recognize the dominant patterns if affected individuals are seen in different generations within the same family, the pattern may be hard to determine when family history is lacking or there is only a single case available. In an autosomal dominant disease, a patient inherits one deleterious copy of a disease-causal gene and either parent may be clinically affected and carry the same defective gene. This variant is usually a missense mutation rather than the loss-of-function mutations typical of recessive diseases. Because it is technically more challenging to identify the causal mutation from the large number of heterozygous mutations in any individual genome, the strategy for sequencing a dominant disease requires more careful selection of samples. Strategies used include sequencing multiple affected and unaffected members in one family or across many families. The data is then analyzed to look for heterozygous mutations that are shared by multiple affected but not unaffected individuals. An example of the success of this strategy is illustrated by the use of exome sequencing to identify *MYH3* mutations shared by four unrelated individuals with Freeman–Sheldon syndrome, a dominantly inherited Mendelian disorder (2). Liu et al. also investigated six unrelated patients with chronic mucocutaneous candidiasis disease (CMCD) by exome sequencing and found heterozygous gain-of-function *STAT1* mutations in four of the studied patients (3).

2.3. X-Linked Inheritance

The typical inheritance pattern for an X-linked recessive disease shows exclusively affected male siblings or cousins in multiple generations within a family. In an affected male, a causal mutation on the X chromosome means that the only copy of this gene is defective. The defective copy is inherited from the unaffected mother who also carries a normal copy of the mutated gene. These variants are usually premature termination or splicing mutations that cause loss of function. For immune-mediated diseases in which the mutant allele has a selective disadvantage, X-linked inheritance can be inferred from a nonrandom X-inactivation against this mutant allele within the immune cells. Such information, gained from either X-inactivation (lyonization) or linkage studies, is very helpful in reducing the potential targets from over 20,000 genes in the human genome down to about 2,000 genes on the X chromosome. For example, Li et al. successfully used X chromosome exome sequencing to identify *MAGT1* loss-of-function mutations as the cause of a new X-linked primary immunodeficiency disease (4).

2.4. Cases Having Unclear or Atypical Genetic Inheritance Patterns

In many cases, the genetic inheritance is not obvious due to small family size or unavailability of family history. Other factors that cloud the genetic inheritance pattern include misassignment of affected status due to genetic or phenotypic heterogeneity, the presence of de novo rather than inherited mutations, and incomplete penetrance due to environmental or other background genetic influences. For such cases, investigators need to consider testing alternative genetic models during data analysis. Furthermore, such cases also underscore the importance of functionally validating any candidate genes gleaned through data analysis.

3. Experimental Design and Selection of Samples

Human genome sequencing has now demonstrated that every individual, healthy or diseased, has a massive number of genetic variants in his or her own genome. The vast majority of these variants are common nucleotide variants or private non-causative variants. The task is to winnow these down to those that are potentially pathogenic and ideally down to the one mutated gene that is responsible for the disease being investigated. The experimental design, especially the selection of the affected/control sample set, can be extremely helpful in distinguishing pathogenically relevant variants from irrelevant ones. Several sample selection strategies to facilitate the downstream data analysis and strategies for filtering variants have been used, which we describe below.

3.1. Sequencing of Parents–Patient Trios

In this strategy, patients and their parents are sequenced. The sequence information from the parents will be used as positive or negative controls, depending on the inheritance model. This approach is especially useful if only a single case is available and can also detect de novo mutations. De novo mutations should be very rare and usually are heterozygous in the patients and not present in both of the parents. Recently, exome sequencing of 20 patients with sporadic autism spectrum disorder and their parents revealed 21 de novo mutations and potential causative de novo events in 4 out of 20 probands (5). Thus, family trio studies allow investigators to generate meaningful results from a relatively small number of sequenced individuals. However, a limitation of this approach is that it may not be suitable for adult-onset genetic diseases, in which parents may be deceased or difficult to locate.

3.2. Sequencing of Multiple Family Members

This approach involves sequencing multiple affected and unaffected individuals from a large family. All affected individuals within the family should share the same causal variant, while the healthy individuals in the same family can be used to filter private benign variants within the family. This sequencing strategy has been successfully used for many genetically heterogeneous disorders, including amyotrophic lateral sclerosis and spinocerebellar ataxias (6–9).

3.3. Sequencing of Unrelated Individuals

In this strategy, multiple unrelated patients who show strikingly similar clinical phenotypes or are diagnosed with the same disorder, are sequenced (10–13). Mutated variants or genes shared by these unrelated affected individuals are more likely to be causal. However, unrelated individuals from different families may have causal variants in different genes if the disorder is not a clear monogenic disease or if phenotypic or genetic heterogeneity exists (14). Therefore, investigators need to be careful in sample selection. Furthermore, they should also consider lowering the stringency when filtering out variants, such as looking for shared variants in most but not all of the affected individuals (15).

4. Selecting the Sequencing Strategy

The recent developments in massively parallel sequencing technologies, coupled with continuous reductions in their cost, allow us to sequence particular targeted genomic regions or even the whole genome to detect genetic variants in the patients. Deciding what to sequence depends on several factors: cost, available sequencing capacity, and bioinformatics support. Here, we will discuss the different options for sequencing strategies.

4.1. Exome Sequencing

The protein-coding regions contained in gene exons, termed the "exome," constitute about 1% of the human genome (2), and it is estimated that mutations involving this small proportion of the genome contribute to most disease-causing mutations (16, 17). Moreover, this approach is now facilitated by affordable commercial kits that allow for efficient isolation or "capture" of DNA comprising the exonic regions of the genome. Thus exome sequencing allows, at a significantly reduced cost, an unbiased identification of the genetic variants that are most likely to cause diseases. The success of this approach has been demonstrated by over one hundred causative disease gene discoveries since its first success reported in 2009 (2, 18). The whole exome capture methods are based on array (2, 17, 19) or in-solution hybridization (20, 21). There are several "off-the-shelf" commercially available whole exome capture products from Agilent Technologies (SureSelect Human All Exon capture kits), Roche-NimbleGen (SeqCap/SeqCap EZ Exome capture kits), and Illumina (TruSeq Exome Enrichment kits). The preferred method depends on several factors including available hybridization equipment (microarray hybridization equipment for array-based hybridization vs. PCR machine for in-solution hybridization), the compatibility of the captured exonic DNA with the downstream sequencing platform, and demand for on-target reads (sensitivity vs. specificity). The known limitation of exome sequencing is that exome capture is unable to cover all the annotated exons due to probe design and capture performance (22). The lack of targeting of the 5′ and 3′ untranslated regions is also a design issue in the current exome capture kits.

4.2. Targeted Capture Sequencing

Prior linkage studies or other information may allow investigators to make educated guesses about which regions or genes may be interesting, thus limiting sequencing efforts to targeted regions of the genome. In a recent study, a custom design of oligonucleotides, which contained 246 genes responsible for human or mouse deafness, was used to identify multiple deleterious mutations in both Arab and Jewish probands who had hearing loss (23). This type of targeted capture sequencing reduces the exome search space and allows screening of more patients with the same amount of sequencing data and cost compared to whole exome sequencing (23–25). The target-enrichment strategies can be very flexible and diverse, depending on the size of the capture regions of interest and the quantity of DNA available. For small regions of interest, traditional PCR and molecular inversion probes (MIP) can be used for target enrichment. For larger regions of interest, investigators can use either hybridization with customized oligonucleotides, similar to what is done in whole exome sequencing, or additional commercial solutions from companies such as Raindance, Halo Genomics, and Febit (26, 27).

4.3. Whole Transcriptome Sequencing (RNA-seq)

Screening the genetic variants that are contained in the human exome can also be accomplished by performing RNA-seq. In this approach, RNA, usually only polyA messenger RNA (mRNA), is converted to cDNA and subjected to DNA sequencing to identify coding variants of expressed genes in particular tissues. Because RNA-seq requires no capture step, this technique is a faster and cheaper approach compared to whole exome capture sequencing (28). In a well-designed validation study, RNA-seq was able to identify 81% of the variants in genes expressed in the source tissue (29). The ability of RNA-seq to identify nucleotide variants depends on the expression level of a particular gene. Thus, it is critical to choose the right tissue or cell type relevant to the clinical traits and to ensure that high-quality RNA from the desired tissue/cell type is prepared. For example, in a recent study, RNA-seq performed on RNA isolated from patients' B cells was successfully used to identify the causal mutation of a gene responsible for familial B cell lymphocytosis with predisposition to chronic lymphocytic leukemia (30).

The major limitation of RNA-seq is that for a loss-of-function mutation such as premature termination resulting in nonsense-mediated decay at the mRNA level, the signal of the genetic variant may be weakened or depleted. In this situation, RNA-seq may only provide indirect evidence of a noncoding mutation in a gene that affects its expression. Thus, for RNA-seq experiments, it is important to include normal controls from matched tissues in order to analyze for missing signals. Furthermore, because RNA-seq generates absolute gene expression counts, the same data set can be analyzed to identify possible effects of any detected mutation on the downstream regulation of transcriptional pathways. This may potentially provide additional mechanistic insights into disease pathogenesis, especially if a gene regulatory protein has been mutated.

4.4. Whole Genome Sequencing

Targeted capture sequencing or transcriptome sequencing focuses on identifying genetic variants that are contained in coding regions or particular genomic loci. By contrast, whole genome sequencing ensures a completely unbiased and thorough investigation of the genomic abnormalities in the patient. The abnormalities detected include genetic variants in noncoding regions (introns, control regions such as promoters and enhancers, and intergenic regions), as well as structural variants including copy number variants and chromosomal rearrangements, which are known to contribute to Mendelian disorders (31). Although the use of whole genome sequencing has been primarily limited by cost, the cost gap between whole genome sequencing and targeted sequencing is quickly decreasing, making the former option increasingly attractive. However, several issues should also be considered before undertaking this approach. First, more robust bioinformatics tools and data analysis pipelines need to be applied to handle the large number of variants found in whole genome sequencing. Moreover, another widely accepted challenge is how to interpret noncoding variants given the lack of solid knowledge as to their biological significance.

5. Data Analysis Strategy for Filtering Data and Prioritizing Mutation Candidates

Once sequencing data of sufficient quality is generated, the sequenced reads need to be aligned to the reference human genome, and the variants are identified based on the alignment output. Multiple open source programs or commercial packages are available to convert the raw sequence reads to variant calls (32, 33). In general, for every sequenced individual, exome sequencing identifies ~20,000 variants (2), and whole genome sequencing identifies ~4,000,000 variants (34). Here we will focus on the major analysis bottleneck that occurs after variant calls are made, which is to establish appropriate filtering criteria to distinguish causal variants from the large amount of benign variants. The key determinants for filtering criteria are the genetic inheritance model and the sequenced samples for individual cases, as discussed below.

5.1. Strategies for Filtering Common and Case-Specific Variants

Filtering strategies are based on several assumptions: First, causal variants of monogenic diseases should be rare. Thus, the known variants contained in public databases (dbSNP (35), the 1000 Genomes project (36), the HapMap project (37)) or normal unrelated control sequencing data (usually in-house shared controls from other sequencing projects) should be excluded. Second, a causal variant usually alters the protein-coding sequence. Thus, synonymous mutations are not likely to be causal variants. Regulatory elements as well as variants that are far away from exon regions are usually also ignored during initial analysis. After applying these basic filters to the exome sequencing data, the list of potential variant candidates usually can be reduced from about 20,000 to several hundred (38). The efficacy of these general filtering steps is limited by the accuracy of the human gene annotation and also the accuracy and the completeness of the databases of common variants. For example, the dbSNP database is routinely used for filtering common variants, but it is well known that some populations are not well represented in dbSNP, and some pathogenic mutations are also found in dbSNP without detailed annotations. Thus, investigators should apply these basic filtering procedures with caution, recognizing their limitations and that they are constantly updated.

After using the universal filters to remove noncoding variants and known common variants, additional variants can be eliminated based upon the expected genetic model of inheritance. As mentioned in the previous section, applying this filtering is case dependent. In general, the recessive model of disease predicts one of two situations: (1) a homozygous variant in the patient and the same heterozygous variant in the parents (usually for families with known consanguinity), or (2) compound heterozygous variants of the same gene in the patient and one of the two variants detected in each parent. In contrast, the dominant model of disease predicts a heterozygous variant in the patient (and possibly in one of the parents). Where there is uncertainty regarding the inheritance model, the stringency of filters may need to be adjusted to allow for potential incomplete penetrance, de novo mutations, and the existence of additional modifier genes. Filtering based on different models might also be considered.

5.2. Prioritization of Disease-Associated Variants

Depending upon the stringency, filtering can reduce the list of possible causal variants to a tractable 10–30 putative variants/genes. These remaining variants can be prioritized based on their functional predictions and prior biological knowledge in several ways:

(a) Loss-of-function variants (nonsense, splicing variants, and small frameshift indels) are more likely to have deleterious effects and to be causative variants in recessive diseases.

(b) Missense variants can be prioritized based on prediction of pathogenicity or evolutionary conservation. A mutation causing an amino acid change that alters the hydrophobicity or structure, and/or that occurs at a highly conserved protein domain, is likely to affect function and contributes to disease. Various programs have been used for these purposes, such as PolyPhen (39), GERP (40), or SIFT (41) for pathogenic prediction of protein alterations and PhyloP (42), PhastCons (43), or SCONE (44) for conservation prediction. These programs are powerful tools but are highly dependent on the quality of our knowledge about those proteins.

(c) Mutations in genes that are selectively expressed in the affected tissues will be more likely associated with disease. For example, genes expressed in hematopoietic cells can be considered to have high potential to be causal genes for immune disorders. The function or pathway knowledge of the candidate genes can also provide clues for the potential relationship between the genotype and phenotypic spectrum.

6. Functional Validation

Advances in sequencing techniques and analysis have dramatically accelerated the process by which variants are discovered and efforts quickly directed to a limited number of likely gene candidates. Nevertheless, success is more likely when involving genes or parts of the genome with which we are familiar. This is because when we already know the normal biological function of a gene, it becomes more straightforward to understand how a variant in that gene could plausibly contribute (or not) to disease pathogenesis. Thus, in cases that do not involve previously well-characterized variants or severe variants in well-characterized genes, determining whether an uncharacterized variant is pathogenic usually also requires determining that gene's biological function. Functional assays used to validate a gene candidate include in vitro cellular assays or animal models that either recapitulate or rescue relevant disease characteristics. A variety of functional assays may be required to establish a convincing causal link between a candidate gene and a clinical phenotype. Overall, this can be expected to be the most rate-limiting step for many studies.

7. Problems and Current Challenges

Although massively parallel sequencing studies have succeeded in identifying many novel gene mutations responsible for Mendelian disorders, success is not always guaranteed. Negative consequences usually result from either technical limitations or analytical failures. Potential technical issues include the following: (1) undercalling of coding variants due to insufficient sequence coverage (i.e., number of individual reads spanning a given variant) or capture sensitivity; (2) causal noncoding or structural variants that are out of the sequence or analysis scope of the available data set; or (3) somatic mosaicism, which will result in missing or undercalling variants in the target tissues/cell types. Analytical failures occur mostly because of incorrect assumptions regarding the genetic model of inheritance, genetic heterogeneity, or genetic misassignment. In situations in which there is no clear candidate at the end of the data analysis, the investigators should double check the quality or coverage depth in the designed capture/sequencing region. If inadequate, more sequencing coverage or an alternative capture approach might be needed to rescue the current missing regions. If quality or coverage depth is adequate, then investigators should reconsider their previous assumptions regarding the inheritance model/phenotype and analysis parameters. In such cases, success might still be achieved if alternative assumptions regarding genetic inheritance assumptions, or different constraints for filtering variants, are applied.

8. IRB and Consent Issues Related to High-Throughput Sequencing

Finally, investigators should be aware that the use of these newer sequencing approaches must be reviewed explicitly by an Institutional Review Board (IRB). Due to the unprecedented scope of genetic information the investigators can expect to obtain, the ethical issues raised by these newer sequencing approaches include the management of the large data sets (both research-related and incidental findings), the sequence data sharing, and the plan to appropriately return genetic findings to subjects, which should be covered by the consent forms. Before sequencing is conducted, the samples must be collected with appropriate written informed consent and IRB approval that is appropriate for sequencing approaches. Therefore, in some situations, previously collected patients might need to be re-consented under an amended consent form. After the sequence data has been collected and analyzed, the results must be returned per the participants' preference, and the genotypic and phenotypic data should be deposited in a data repository such as NCBI dbGaP database (45) to meet data sharing and confidentiality requirements.

9. Conclusion

This chapter provides a brief review of the design and analysis of massively parallel sequencing projects to identify causal mutations in human immune disorders. Careful planning at all steps—from selecting what patients and controls should be sequenced, choosing the most appropriate sequencing strategy, and determining what kind of filtering process should be applied to the sequence data sets—is critical to increase the likelihood of success for the sequencing project. Researchers have only had a few years' experience with this incredible methodology to rapidly discover the genetic disease-related genes, and we expect to see an increasing number of successful discoveries in the foreseeable future.

Acknowledgments

We thank Bernice Lo and Helen Matthews for critically reading the manuscript. This work was supported by the Intramural Research Program of the National Institutes of Health, the National Institute of Allergy and Infectious Diseases.

References

1. Bolze A et al (2010) Whole-exome-sequencing-based discovery of human FADD deficiency. Am J Hum Genet 87:873–881
2. Ng SB et al (2009) Targeted capture and massively parallel sequencing of 12 human exomes. Nature 461:272–276
3. Liu L et al (2011) Gain-of-function human STAT1 mutations impair IL-17 immunity and underlie chronic mucocutaneous candidiasis. J Exp Med 208:1635–1648
4. Li FY et al (2011) Second messenger role for Mg2+ revealed by human T-cell immunodeficiency. Nature 475:471–476
5. O'Roak BJ et al (2011) Exome sequencing in sporadic autism spectrum disorders identifies severe de novo mutations. Nat Genet 43: 585–589
6. Johnson JO et al (2010) Exome sequencing reveals VCP mutations as a cause of familial ALS. Neuron 68:857–864
7. Wang JL et al (2010) TGM6 identified as a novel causative gene of spinocerebellar ataxias using exome sequencing. Brain 133:3510–3518
8. Musunuru K et al (2010) Exome sequencing, ANGPTL3 mutations, and familial combined hypolipidemia. N Engl J Med 363:2220–2227
9. Krawitz PM et al (2010) Identity-by-descent filtering of exome sequence data identifies PIGV mutations in hyperphosphatasia mental retardation syndrome. Nat Genet 42:827–829
10. Ng SB et al (2010) Exome sequencing identifies the cause of a mendelian disorder. Nat Genet 42:30–35
11. Lalonde E et al (2010) Unexpected allelic heterogeneity and spectrum of mutations in Fowler syndrome revealed by next-generation exome sequencing. Hum Mutat 31:918–923
12. Pierce SB et al (2010) Mutations in the DBP-deficiency protein HSD17B4 cause ovarian dysgenesis, hearing loss, and ataxia of Perrault Syndrome. Am J Hum Genet 87:282–288
13. Hoischen A et al (2010) De novo mutations of SETBP1 cause Schinzel-Giedion syndrome. Nat Genet 42:483–485
14. Gilissen C et al (2010) Exome sequencing identifies WDR35 variants involved in Sensenbrenner syndrome. Am J Hum Genet 87:418–423

15. Ng SB et al (2010) Exome sequencing identifies MLL2 mutations as a cause of Kabuki syndrome. Nat Genet 42:790–793
16. Botstein D, Risch N (2003) Discovering genotypes underlying human phenotypes: past successes for mendelian disease, future approaches for complex disease. Nat Genet 33(Suppl):228–237
17. Choi M et al (2009) Genetic diagnosis by whole exome capture and massively parallel DNA sequencing. Proc Natl Acad Sci USA 106:19096–19101
18. Ku CS et al (2011) Revisiting Mendelian disorders through exome sequencing. Hum Genet 129:351–370
19. Okou DT et al (2007) Microarray-based genomic selection for high-throughput resequencing. Nat Methods 4:907–909
20. Gnirke A et al (2009) Solution hybrid selection with ultra-long oligonucleotides for massively parallel targeted sequencing. Nat Biotechnol 27:182–189
21. Bainbridge MN et al (2010) Whole exome capture in solution with 3 Gbp of data. Genome Biol 11:R62
22. Sulonen AM et al (2011) Comparison of solution-based exome capture methods for next generation sequencing. Genome Biol 12:R94
23. Brownstein Z et al (2011) Targeted genomic capture and massively parallel sequencing to identify genes for hereditary hearing loss in middle eastern families. Genome Biol 12:R89
24. Nikopoulos K et al (2010) Next-generation sequencing of a 40 Mb linkage interval reveals TSPAN12 mutations in patients with familial exudative vitreoretinopathy. Am J Hum Genet 86:240–247
25. Rehman AU et al (2010) Targeted capture and next-generation sequencing identifies C9orf75, encoding taperin, as the mutated gene in nonsyndromic deafness DFNB79. Am J Hum Genet 86:378–388
26. Tewhey R et al (2009) Microdroplet-based PCR enrichment for large-scale targeted sequencing. Nat Biotechnol 27:1025–1031
27. Johansson H et al (2011) Targeted resequencing of candidate genes using selector probes. Nucleic Acids Res 39:e8
28. Chepelev I et al (2009) Detection of single nucleotide variations in expressed exons of the human genome using RNA-Seq. Nucleic Acids Res 37:e106
29. Cirulli ET et al (2010) Screening the human exome: a comparison of whole genome and whole transcriptome sequencing. Genome Biol 11:R57
30. Snow AL et al (2012) Congenital B cell lymphocytosis explained by novel germline CARD11 mutations. J Exp Med 2012 Nov 5 epub ahead of print.
31. Lupski JR et al (2010) Whole-genome sequencing in a patient with Charcot-Marie-Tooth neuropathy. N Engl J Med 362:1181–1191
32. Horner DS et al (2010) Bioinformatics approaches for genomics and post genomics applications of next-generation sequencing. Brief Bioinform 11:181–197
33. Meyerson M et al (2010) Advances in understanding cancer genomes through second-generation sequencing. Nat Rev Genet 11:685–696
34. Bentley DR et al (2008) Accurate whole human genome sequencing using reversible terminator chemistry. Nature 456:53–59
35. Sherry ST et al (2001) dbSNP: the NCBI database of genetic variation. Nucleic Acids Res 29:308–311
36. Consortium, T. G. P (2010) A map of human genome variation from population-scale sequencing. Nature 467:1061–1073
37. Altshuler DM et al (2010) Integrating common and rare genetic variation in diverse human populations. Nature 467:52–58
38. Stitziel NO et al (2011) Computational and statistical approaches to analyzing variants identified by exome sequencing. Genome Biol 12:227
39. Adzhubei IA et al (2010) A method and server for predicting damaging missense mutations. Nat Methods 7:248–249
40. Cooper GM et al (2010) Single-nucleotide evolutionary constraint scores highlight disease-causing mutations. Nat Methods 7:250–251
41. Ng PC, Henikoff S (2003) SIFT: predicting amino acid changes that affect protein function. Nucleic Acids Res 31:3812–3814
42. Pollard KS et al (2010) Detection of nonneutral substitution rates on mammalian phylogenies. Genome Res 20:110–121
43. Siepel A et al (2005) Evolutionarily conserved elements in vertebrate, insect, worm, and yeast genomes. Genome Res 15:1034–1050
44. Asthana S et al (2007) Analysis of sequence conservation at nucleotide resolution. PLoS Comput Biol 3:e254
45. Mailman MD et al (2007) The NCBI dbGaP database of genotypes and phenotypes. Nat Genet 39:1181–1186

Chapter 15

Flow Cytometric Measurement of SLAM-Associated Protein and X-Linked Inhibitor of Apoptosis

Rebecca A. Marsh, Jack J. Bleesing, and Alexandra H. Filipovich

Abstract

Flow cytometry is a valuable tool for the detection and characterization of proteins expressed by individual cells. Flow cytometry can be used to measure cell expression of 2 intracellular proteins that are involved in the regulation of immune homeostasis, SLAM-associated protein (SAP) and X-linked inhibitor of apoptosis (XIAP). These proteins are defective in patients with the immune deficiency X-linked lymphoproliferative disease (XLP), due to mutations in the *SH2D1A* and *XIAP/BIRC4* genes, respectively (Coffey et al. Nat Genet 20:129-135 1998; Nichols et al. Proc Natl Acad Sci U S A 95:13765-13770, 1998; Sayos et al. Nature 395:462-469, 1998; Rigaud et al. Nature 444:110-114, 2006). This procedure describes a technique that can be efficiently used to detect SAP and XIAP by flow cytometry.

Key words: X-linked Lymphoproliferative Disease (XLP), SLAM-Associated Protein (SAP), X-linked Inhibitor of Apoptosis (XIAP)

1. Introduction

X-linked lymphoproliferative disease (XLP) is a life-threatening immune deficiency that is caused by defects in SLAM-associated protein (SAP) or X-linked inhibitor of apoptosis (XIAP), due to mutations in the *SH2D1A* or *XIAP/BIRC4* genes, respectively (1–4). Flow cytometry can be used to measure the intracellular expression of SAP and XIAP in human leukocytes. This can be useful to researchers and allows clinicians to quickly screen patients for XLP. SAP is expressed primarily in NK cells and T cells, and XIAP is expressed in all hematopoietic cells (4–7). Here we describe techniques that can be used to detect SAP or XIAP in primary human lymphocytes using flow cytometry (5–8). These procedures can be adapted for use with cultured cells, immortalized cell lines, or tumor cell lines.

Andrew L. Snow and Michael J. Lenardo (eds.), *Immune Homeostasis: Methods and Protocols*, Methods in Molecular Biology, vol. 979, DOI 10.1007/978-1-62703-290-2_15, © Springer Science+Business Media New York 2013

2. Materials

2.1. Specimens

Whole human blood (>200 μl) (see Note 1). Peripheral blood samples should be anti-coagulated (see Note 2).

2.2. Equipment

1. Pipettor and pipet tips, for 2–200 μl.
2. Dispenser or pipets for dispensing 2 ml.
3. Centrifuge with swinging bucket rotor and adapters for 12×75 mm polystyrene tubes.
4. Vortex mixer.
5. Flow cytometer and analysis software of choice.
6. 12×75 mm nonsterile polystyrene round bottom test tubes.

2.3. Reagents

1. Intraprep permeabilization kit (Beckman Coulter).
2. 1% paraformaldehyde (or other appropriate user-preferred fixative).
3. 1× Phosphate buffered saline (PBS).

2.4. Antibodies

As an illustration, we recommend the following collection of antibodies used for the detection of SAP in human lymphocytes (T, B, and NK cells) (see Note 3):

1. Fluorochrome-conjugated mouse anti-human CD3 (APC-, BD Biosciences, #340661).
2. Fluorochrome conjugated mouse anti-human CD8 (PerCP-, BD Biosciences, #340693).
3. Fluorochrome conjugated mouse anti-human CD56 (PE-, BD Biosciences, #340685).
4. Rat IgG_1—purified (BD Biosciences, #555839).
5. Rat anti-human SAP antibody that has been validated for flow cytometric use (e.g., clone KST-3; see Note 4). For this clone, a FITC-conjugated goat anti-rat IgG_1 antibody is used for the detection of anti-SAP or isotype control antibody (BD Biosciences, catalog # 553892).

Recommended antibodies used for the detection of XIAP in human lymphocytes:

1. Fluorochrome conjugated mouse anti-human CD3 (PerCP-, BD Biosciences, #340663).
2. Fluorochrome conjugated mouse anti-human CD4 (PacBlue-, BD Biosciences, #558116).
3. Fluorochrome conjugated mouse anti-human CD56 (APC-, Beckman Coulter, #IM2474).

4. Fluorochrome conjugated mouse anti-human CD19 (FITC-, BD Biosciences, #340719).
5. Mouse IgG_1—purified (BD Biosciences, #349040).
6. Mouse anti-human XIAP antibody that has been validated for flow cytometric use (e.g., clone 48) (see Note 5). For this clone, a PE-conjugated goat anti-mouse IgG_1 antibody is used for detection of anti-XIAP or isotype control antibody (BD Biosciences, catalog #550589).

3. Methods

3.1. Intracellular Staining for SAP

1. Add 100 μl whole blood to each of 2 tubes, labeled "isotype" and "SAP" (see Note 6).
2. Add appropriate amounts of fluorochrome-conjugated mouse anti-human CD3, mouse anti-human CD8, and mouse anti-human CD56 to each tube.
3. Vortex and incubate in the dark at room temperature (RT) for 15 min.
4. Add 100 μl IntraPrep Reagent 1 (the fixative) from the IntraPrep Reagent kit, vortex, and incubate in the dark at RT for 15 min.
5. Add 2 ml PBS and centrifuge 5 min at 300 × *g*.
6. Aspirate the supernatant and resuspend the pellet COMPLETELY by vortexing (see Note 8).
7. Add 100 μl Reagent #2 (the permeabilizer). Do not mix.
8. Incubate in the dark at RT for 5 min.
9. Add an appropriate amount of the rat anti-SAP antibody and the rat IgG_1 isotype control to the "SAP" and "isotype" tubes, respectively, and vortex gently (see Note 7). Incubate in the dark at RT for 20 min. Add 2 ml PBS and centrifuge 5 min at 300 × *g*.
10. Aspirate the supernatant and add an appropriate amount of FITC-conjugated anti-rat antibody in 100 μl PBS.
11. Incubate in the dark at RT for 20 min.
12. Add 2 ml PBS and centrifuge 5 min at 300 × *g*.
13. Aspirate the supernatant and then resuspend the pellet in 200 μl 1% paraformaldehyde and vortex gently (or use an appropriate amount of a user-preferred fixative).
14. Run immediately on flow cytometer or store at 4°C in the dark until analysis.

3.2. Flow Cytometric Analysis of Intracellular SAP Expression

1. Using the flow cytometer and software with which you are familiar, first set a gate on the lymphocyte population, based on forward and side light scatter characteristics.
2. Starting with gated lymphocytes, draw a 2-color dot plot to differentiate T cells from NK cells by plotting $CD3^+$ against $CD56^+$.
3. Create gates for the $CD3^+$ T cell population and the $CD56^+$ $CD3^-$ NK cell populations.
4. Using the T cell gate, draw another dot plot to differentiate $CD8^+$ T cells from $CD8^-$ T cells by plotting $CD3^+$ against $CD8^+$. Create gates on the $CD8^+$ and $CD8^-$ populations.
5. Create histograms to visualize SAP expression within NK cells, $CD8^+$ T cells, and $CD8^-$ T cells. Compare the SAP histogram to the histogram generated from the isotype control, which serves as a negative control for background/nonspecific fluorescence. Set a marker at the intersection of the two peaks of each histogram to determine the percentage of cells that have detectable SAP. See Fig. 1 for sample gating strategy and analysis of SAP expression.

3.3. Intracellular Staining for XIAP

1. Add 100 μl whole blood to each of 2 tubes, labeled “isotype” and “XIAP” (see Note 9).
2. Add 150 μl IntraPrep Reagent 1 into both tubes, vortex, and incubate at room temperature for 15 min.
3. Add 2 ml PBS and centrifuge 5 min at 300 × *g*.
4. Aspirate the supernatant and resuspend COMPLETELY in residual liquid.
5. Add 150 μl IntraPrep Reagent 2 to each tube. Do not vortex.
6. Make a 1:10 dilution of the anti-XIAP antibody in PBS. Add 10 μl of diluted XIAP antibody reagent to the “XIAP” tube and 5 μl Ms IgG_1 isotype control to the “isotype” tube. Vortex very gently and incubate in the dark at RT for 20 min.
7. Add 2 ml PBS and centrifuge 5 min at 300 × *g*.
8. Aspirate the supernatant and resuspend cells. Add an appropriate amount of the PE-conjugated anti-mouse reagent to all tubes.
9. Incubate in the dark at RT for 20 min.
10. Add 2 ml PBS and centrifuge 5 min at 300 × *g*.
11. Aspirate the supernatant and resuspend cells in ~100 μl PBS. Add 10 μl purified Ms IgG1 to all tubes to bind any residual anti-mouse antibody.
12. Incubate 10 min in the dark at RT.

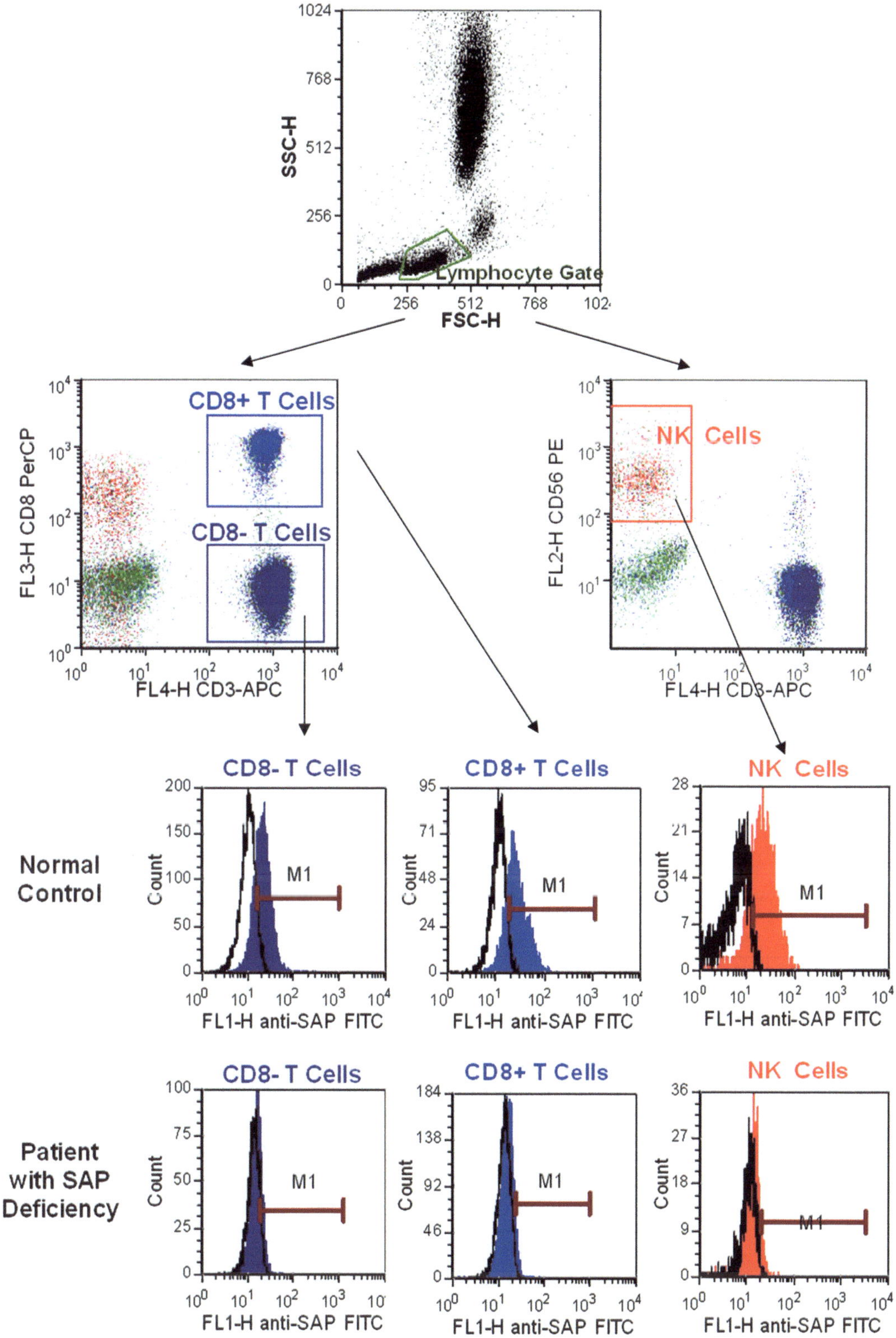

Fig.1. Sample gating strategy and analysis of SAP expression by human peripheral blood T cells and NK cells. Examples of SAP expression in peripheral human blood lymphocytes from a normal control (*top row* of histograms) and from a patient with XLP due to SAP deficiency (*bottom row* of histograms) is shown.

13. Add appropriate amounts of fluorochrome conjugated anti-CD19, anti-CD3, anti-CD56, and anti-CD4. Incubate in the dark at RT for 15 min.
14. Add 2 ml PBS and centrifuge 5 min at $300\times g$.
15. Aspirate the supernatant and resuspend in 200 μl 1% paraformaldehyde (or use an appropriate amount of a user-preferred fixative).
16. Run immediately on flow cytometer or store at 4°C in the dark until analysis.

3.4. Flow Cytometric Analysis of Intracellular XIAP Expression

1. Draw a gate to designate the lymphocyte population, based on forward and side light scatter characteristics.
2. Create a 2-color dot plot to differentiate $CD4^+$ T cells from $CD4^-$ T cells by plotting $CD3^+$ against $CD4^+$. Create gates on the $CD3^+CD4^+$ and $CD3^+CD4^-$ T cell populations.
3. Create a dot plot to identify NK cells by plotting $CD3^+$ against $CD56^+$. Create a gate for the $CD56^+CD3^-$ NK cell population.
4. Create a dot plot to identify B cells by plotting $CD19^+$ against $CD3^+$. Create a gate on the $CD19^+CD3^-$ B cell population.
5. Create histograms to visualize XIAP expression within gated $CD4^+$ T cells, $CD4^-$ T cells, NK cells, and B cells. Compare the XIAP histogram to the isotype control, which serves as a negative control for background/nonspecific fluorescence. Set a marker at the intersection of the two peaks of each histogram to determine the percentage of cells with detectable XIAP expression. See Fig. 2 for sample gating strategy and analysis of XIAP expression.

4. Notes

1. These techniques can be modified to measure SAP or XIAP in Ficoll-isolated peripheral blood mononuclear cells (see Chapter 2), cultured cells, or cell lines. These cell sources should be in a single cell suspension at a cell concentration appropriate for flow cytometry work, within a media or buffer appropriate for flow cytometry work.
2. In our experience, it is preferable to use heparin or ACD as anticoagulants when measuring SAP, and EDTA is preferred when measuring XIAP.
3. The method described below can be used to detect SAP and XIAP within specific lymphocyte subsets using cell surface antibody staining to differentiate T cells, NK cells, and B cells. We illustrate the use of 4- and 5-color flow cytometric analysis

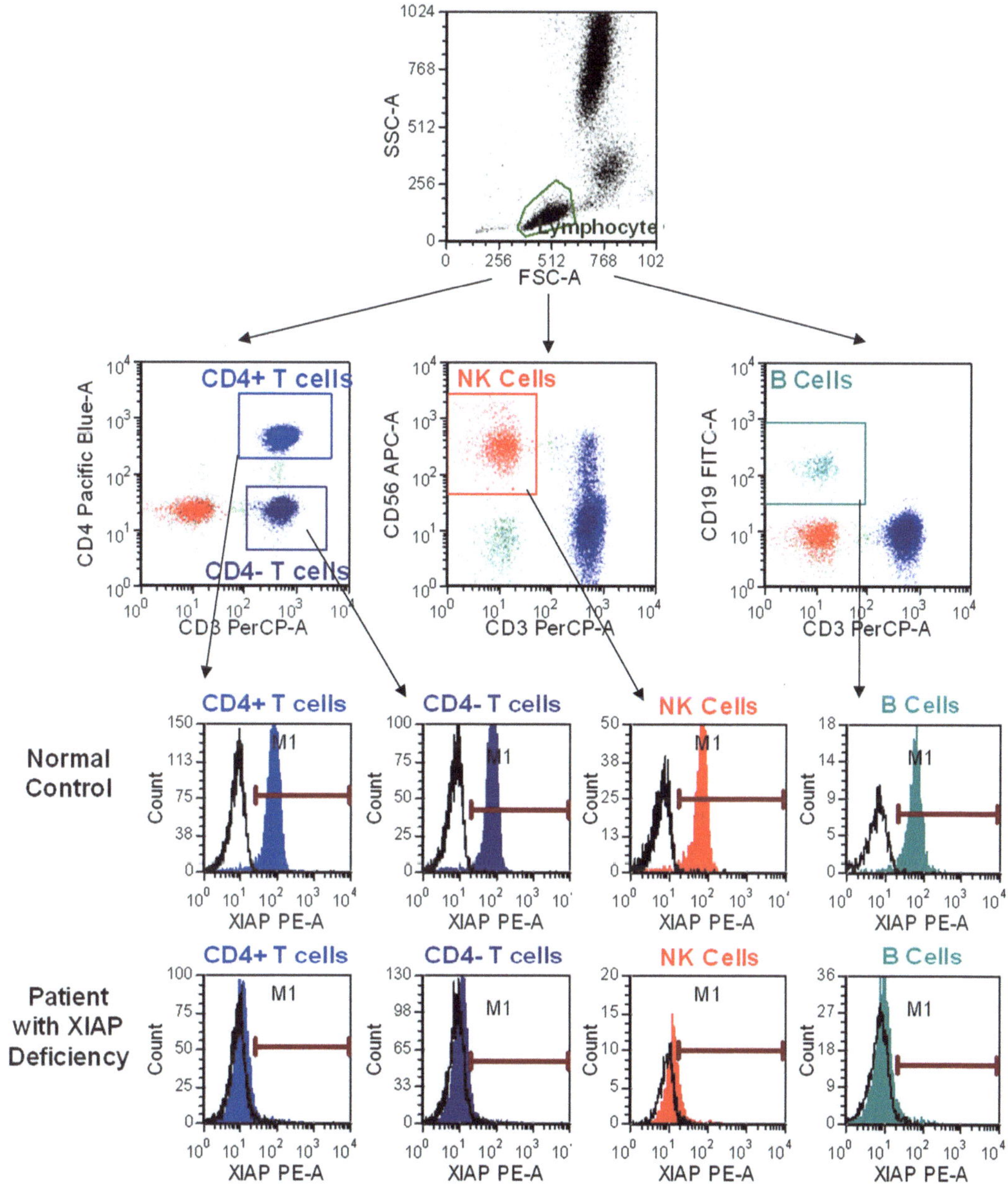

Fig. 2. Sample gating strategy and analysis of XIAP expression by human peripheral blood T cells, NK cells, and B cells. Examples of XIAP expression in peripheral human blood lymphocytes from a normal control (*top row* of histograms) and from a patient with XLP due to XIAP deficiency (*bottom row* of histograms) is shown.

using combinations of the fluorochromes fluorescein isothiocyanate (FITC), R-phycoerythrin (PE), allophycocyanin (APC), peridinin chlorophyll protein (PerCP), and Pacific Blue (PacBlue). The user can adapt the surface staining antigens and fluorochromes based on their needs and flow cytometer capabilities.

4. For the purpose of this protocol, we demonstrate the detection of SAP using antibody that was produced and purified from the hybridoma clone KST-3, a gift from Dr. Hirokazu Kanegane. Antibody production and purification was performed by East Coast Biologics.
5. For the purpose of this protocol, we demonstrate the detection of XIAP using antibody obtained from BD Biosciences (clone 48, catalog #610762).
6. Patients with XLP who are ill with hemophagocytic lymphohistiocytosis may have significant expansion of T cell populations. In these cases, it may be helpful to decrease the amount of blood used. Cells may also appear to have nonspecific staining. Washing of the blood may help to resolve this.
7. If a murine anti-SAP antibody is used, it must be directly conjugated to an appropriate fluorochrome. Alternatively, the surface staining may be performed at the end of the assay if a non-conjugated murine anti-SAP antibody is used, followed by a fluorochrome-conjugated anti-mouse antibody.
8. Red cell lysis will not be complete if the cells are not completely resuspended after incubation with IntraPrep Reagent 1 before the addition of Reagent 2. Take care to vortex and manually mix until the red cell pellet is completely resuspended.
9. As samples age, detection of XIAP decreases. Analysis of XIAP in peripheral human blood lymphocytes should be performed within 24 h of the blood being drawn.

References

1. Coffey AJ, Brooksbank RA, Brandau O, Oohashi T, Howell GR, Bye JM, Cahn AP, Durham J, Heath P, Wray P, Pavitt R, Wilkinson J, Leversha M, Huckle E, Shaw-Smith CJ, Dunham A, Rhodes S, Schuster V, Porta G, Yin L, Serafini P, Sylla B, Zollo M, Franco B, Bolino A, Seri M, Lanyi A, Davis JR, Webster D, Harris A, Lenoir G, de St Basile G, Jones A, Behloradsky BH, Achatz H, Murken J, Fassler R, Sumegi J, Romeo G, Vaudin M, Ross MT, Meindl A, Bentley DR (1998) Host response to EBV infection in X-linked lymphoproliferative disease results from mutations in an SH2-domain encoding gene. Nat Genet 20:129–135
2. Nichols KE, Harkin DP, Levitz S, Krainer M, Kolquist KA, Genovese C, Bernard A, Ferguson M, Zuo L, Snyder E, Buckler AJ, Wise C, Ashley J, Lovett M, Valentine MB, Look AT, Gerald W, Housman DE, Haber DA (1998) Inactivating mutations in an SH2 domain-encoding gene in X-linked lymphoproliferative syndrome. Proc Natl Acad Sci USA 95:13765–13770
3. Sayos J, Wu C, Morra M, Wang N, Zhang X, Allen D, van Schaik S, Notarangelo L, Geha R, Roncarolo MG, Oettgen H, De Vries JE, Aversa G, Terhorst C (1998) The X-linked lymphoproliferative-disease gene product SAP regulates signals induced through the co-receptor SLAM. Nature 395:462–469
4. Rigaud S, Fondaneche MC, Lambert N, Pasquier B, Mateo V, Soulas P, Galicier L, Le Deist F, Rieux-Laucat F, Revy P, Fischer A, de Saint Basile G, Latour S (2006) XIAP deficiency in humans causes an X-linked lymphoproliferative syndrome. Nature 444: 110–114
5. Shinozaki K, Kanegane H, Matsukura H, Sumazaki R, Tsuchida M, Makita M, Kimoto Y, Kanai R, Tsumura K, Kondoh T, Moriuchi H, Miyawaki T (2002) Activation-dependent T cell expression of the X-linked lymphoproliferative disease gene product SLAM-associated protein and its assessment for patient detection. Int Immunol 14:1215–1223

6. Marsh RA, Villanueva J, Zhang K, Snow AL, Su HC, Madden L, Mody R, Kitchen B, Marmer D, Jordan MB, Risma KA, Filipovich AH, Bleesing JJ (2009) A rapid flow cytometric screening test for X-linked lymphoproliferative disease due to XIAP deficiency. Cytometry B Clin Cytom 76:334–344
7. Tabata Y, Villanueva J, Lee SM, Zhang K, Kanegane H, Miyawaki T, Sumegi J, Filipovich AH (2005) Rapid detection of intracellular SH2D1A protein in cytotoxic lymphocytes from patients with X-linked lymphoproliferative disease and their family members. Blood 105:3066–3071
8. Zhao M, Kanegane H, Kobayashi C, Nakazawa Y, Ishii E, Kasai M, Terui K, Gocho Y, Imai K, Kiyasu J, Nonoyama S, Miyawaki T (2011) Early and rapid detection of X-linked lymphoproliferative syndrome with SH2D1A mutations by flow cytometry. Cytometry B Clin Cytom 80:8–13

Chapter 16

In Vitro Suppression Assay for Functional Assessment of Human Regulatory T Cells

Dat Q. Tran

Abstract

FOXP3+ regulatory T cells (Tregs) play an important role in controlling immune activation, maintenance of homeostasis and the prevention of autoimmunity. Much effort has been focused in assessing their potential defects in certain human diseases and in developing potential Treg immunotherapy to cure autoimmune conditions. One cardinal feature of Tregs is their ability to suppress the activation and proliferation of effector T cells. The in vitro suppression assay is a convenient and relatively reliable method to quantify their suppressive function and validate their identity as true Tregs. This protocol describes a reliable, nonradioactive in vitro suppression assay to assess the immunosuppressive property of human Tregs in inhibiting proliferation of dye-labeled responder T cells using non-irradiated, HLA-DR+ antigen presenting cells and soluble anti-CD3 antibodies as stimuli.

Key words: FOXP3, Regulatory T cells, Tregs, Suppression assay, Autoimmunity, Immunosuppression

1. Introduction

Regulatory T cells (CD4+FOXP3+, Tregs) are critical for the establishment of self-tolerance early in life and its maintenance throughout adulthood (1). The transcription factor FOXP3 is essential for the development of Tregs (2). Mutations in the *FOXP3* gene result in a rare human autoimmune condition called immune dysregulation polyendocrinopathy enteropathy X-linked disease (IPEX), which manifests during infancy with systemic autoimmunity affecting the intestine, endocrine, skin, and hematopoetic systems due to deficiency or dysfunction in Tregs (3). Other autoimmune and transplant-related conditions have been associated with a deficiency or dysfunction of Tregs (4, 5). However, these results have been challenged due to issues with the purity of isolated Tregs and reliability of in vitro functional assays.

Earlier studies isolated Tregs based on the correlative high expression of the surface marker CD25, the alpha chain of the

Andrew L. Snow and Michael J. Lenardo (eds.), *Immune Homeostasis: Methods and Protocols*, Methods in Molecular Biology, vol. 979, DOI 10.1007/978-1-62703-290-2_16, © Springer Science+Business Media New York 2013

interleukin-2 (IL-2) receptor. Since CD25 is also upregulated on activated T cells, the purity of Tregs becomes problematic in disease due to contamination of $CD4^+CD25^+FOXP3^-$ conventional T cells. The recent discovery that Tregs also express low to undetectable levels of CD127, the alpha chain of IL-7 receptor, allows for greater yield and purity of Tregs by minimizing contamination of activated conventional T cells.

Another issue of assessing the function of Tregs is the variability and reliability of in vitro functional assays. Considering that the mechanisms of Treg-mediated suppression remain unresolved, there is no gold standard to validate these assays except for the inhibition of T cell activation and proliferation in the presence of Tregs (6). A controversy exists as to which cell the Tregs target in mediating their suppression. In the assay that uses either plate bound or bead-conjugated anti-CD3/CD28 antibodies (Abs) for stimulation in the absence of antigen presenting cells (APCs), the presumption is that the Tregs target the responder T cells directly to suppress their activation and proliferation. Since Tregs express high levels of the high-affinity IL-2 receptor, one potential suppressive effect in these assays stems from IL-2 consumption, particularly at ratios of 1:1 and 1:2 Treg to responder T cells. Another potential problem is competitive interaction with the stimulus. Because a cardinal feature of Tregs is anergy, defined as the lack of proliferation and cytokine production upon T cell receptor (TCR) activation, inhibition of responder T cell activation could occur at limiting concentrations of stimulating reagents due to competitive inhibition by the Tregs. Assays using APCs, typically T-cell depleted splenocytes in murine system or PBMC in human system, appear to be more physiologic but are still susceptible to in vitro artificial suppressive responses. Evidence suggests that Tregs target APCs or DCs to mediate their suppression of immune activation in addition to any direct effect on T cells (7–9). Until the dominant mechanism of Treg-mediated suppression is clarified, it is important to select an assay that is consistent and reproducible with minimal nonspecific responses. This protocol describes in detail how to set up and interpret an in vitro Treg suppression assay that utilizes non-irradiated $HLA\text{-}DR^+$ cells as APCs and $CD4^+CD25^-$ responder T cells. The readout of suppression will be proliferation of responder cells based on their dilution of carboxyfluorescein succinimidyl ester (CFSE), although tritiated thymidine (^3H-TdR) incorporation can be substituted.

2. Materials

2.1. Cultureware and Media

1. 96-well clear, sterile polystyrene flat-bottom culture plates (Corning, Nunc or equivalent).
2. Multichannel pipette that can hold three pipette tips (200 μl volume).

3. Eppendorf repeater pipette or equivalent.
4. Complete culture medium: RPMI-1640 supplemented with 10% heat-inactivated Fetal Bovine Serum (FBS), 100 U/ml penicillin, 100 μg/ml streptomycin, 2 mM L-glutamine, 10 mM HEPES, 0.1 mM nonessential amino acids, and 1 mM sodium pyruvate.
5. Ficoll-Paque or equivalent for isolation of PBMC from whole blood or buffy coat.
6. 5 ml polystyrene FACS tubes (Becton Dickinson).

2.2. Antibodies and Reagents

1. Anti-CD3 Abs (i.e., OKT3 or UCHT1 clone)
2. CFSE (5 mM stock concentration; kit available from Invitrogen).
3. Fluorochrome-conjugated antibodies for flow cytometry: anti-CD4, anti-CD25, and anti-CD127.
4. Anti-CD4 microbeads for magnetic sorting (see Note 1).
5. FOXP3 intracellular staining kit (eBioscience).
6. Alexa-Fluor 488 conjugated anti-FOXP3 Abs (clone 206D, 259D, or 236A/E7).
7. Sterile FACS staining buffer: 1× PBS (Ca/Mg^{++} free), 1% w/v bovine serum albumin (BSA).
8. Sterile FACS sorting buffer: 1× PBS, 1 mM EDTA, 2 mM HEPES, 1% FBS).

2.3. Equipment

1. Magnetic cell sorter (e.g., Miltenyi Biotec or equivalent) and columns.
2. Fluorescence-activated cell sorter (e.g., FACSAria).
3. Flow cytometer and appropriate analysis software.

3. Methods

3.1. Isolation of Responders ($CD4^+CD25^-$ T Cells) and Tregs ($CD4^+CD127^-CD25^{high}$) by FACS (see Note 2)

1. Isolate PBMC from either whole blood or buffy coat with density gradient centrifugation using Ficoll-Paque or equivalent per manufacturer's protocol (see Chapter 2).
2. Isolate $CD4^+$ T cells from PBMC using anti-CD4 magnetic beads and run the sample through a magnetic column per manufacturer's protocol (see Note 2). After isolation of $CD4^+$ T cells, save the $CD4^-$ fraction to obtain APCs (see Note 3).
3. Resuspend cells in 50 μl of sterile FACS buffer per 10^7 total cells. Label the $CD4^+$ cells by adding the appropriate amount of fluorochrome-conjugated Abs to detect CD4, CD25 and CD127 (see Note 4). Incubate cells at 4°C for 20–30 min.

4. After staining, wash the cells to remove the unlabeled antibodies by adding 10× the volume of FACS buffer and centrifuging at $300 \times g$ for 5–10 min. Remove the supernatant and resuspend the cells in FACS sorting buffer at a concentration of $10–20 \times 10^6$ cells per ml, depending on the preference of the sorting facility.
5. Sort the cells on the FACS machine (see Note 5). Draw gates for the $CD4^+CD127^+$ T cell and $CD4^+CD127^-$ T cell populations. In the $CD127^+$ gate, sort for responders by gating on $CD4^+CD25^-$. In the $CD127^-$ gate, sort for Tregs by gating on $CD4^+CD25^{high}$ cells, which should be restricted to 2–4% of the $CD4^+$ population with the highest expression of CD25 to minimize $CD25^{low}FOXP3^-$ contamination (see Note 6).
6. After completion of sorting, determine the post-sort purity based on FOXP3 expression (20,000 responders, 20,000 Tregs) using the FOXP3 staining kit and Alexa-Fluor 488 conjugated anti-FOXP3 Abs (clone 206D, 259D, or 236A/E7) (see Note 7).

3.2. Isolation of HLA-DR+ APCs (see Note 8)

1. From PBMC, deplete $CD4^+$ cells with anti-CD4 magnetic beads to remove a significant subset of Tregs that expresses HLA-DR.
2. Add HLA-DR beads to the CD4-depleted fraction to isolate $HLA\text{-}DR^+$ APCs by magnetic cell sorting, according to the manufacturer's instructions (see Note 9).
3. To achieve physiologic conditions and to avoid perturbation of potential vital processes involved in the suppressive mechanism, the APCs are not irradiated. Whether the APCs survive and proliferate during the 4–5 day culture will not affect the detection of responder proliferation because they will be identified based on CFSE dilution. In this culture condition, most of the APCs will not survive by day 4–5 of culture so that ^3H-TdR incorporation could be performed with minimal background.

3.3. Labeling Responders with CFSE (see Note 10)

1. Determine how many responders are required for the assay and use 25–50% more cell numbers for the CFSE staining to compensate for potential cell loss from the washing process. A typical assay uses 25,000–50,000 responders per well in triplicates.
2. It is important to achieve optimal, bright CFSE staining so that at least 3–5 cell divisions of the CFSE population have not dropped to the level of the unstained population (see Note 11).
3. Place the appropriate cell number in a 50 ml sterile conical centrifuge tube and pellet the cells by centrifugation at $300 \times g$ for 5–10 min.
4. Decant the supernatant and resuspend the cells in 37°C FACS buffer at a final concentration of 10^6/ml. If the cells were in

serum-containing medium, they should be washed twice with FACS buffer to remove the serum and avoid quenching the CFSE fluorescence.

5. Pipet the appropriate volume of the 5 mM stock CFSE solution to achieve a final working concentration of 10 μM, which should allow for detection of the CFSE-labeled cells from the unlabeled cells after 3–5 divisions during the 3–5 day culture.
6. Incubate the cells at 37°C for 10 min to allow for absorption of the CFSE. Afterward, stop the staining process by adding refrigerated complete medium to the 40 ml mark on the tube.
7. Mix the solution by rotating the tube back and forth several times, then centrifuge at $300 \times g$ (4°C) for 10 min.
8. Decant the supernatant and wash the cells two more times with 30 ml complete medium. After the last wash, resuspend the cells in complete medium to give an estimated cell concentration of 2×10^6/ml.
9. Count the cells to confirm the actual cell number and add more complete media if needed to achieve a final concentration of 10^6/ml. For confirmation of staining, take a small aliquot of unlabeled and labeled cells to the flow cytometer to determine the intensity of CFSE staining.

3.4. Set Up of T Cell Suppression Assay (see Note 12)

1. Resuspend all T cell populations in complete medium at a cell concentration of 1×10^6 cells/ml. Resuspend APCs at 0.5×10^6 cells/ml.
2. Label a 96-well plate (A-H for rows and 1–12 for columns). Use each row across for each cell population that is being tested for its suppressive function (Fig. 1a). Row A is reserved for CFSE-labeled responders only. Row B contains a negative control (non-Tregs) consisting of CFSE-labeled responders cocultured with decreasing numbers of unlabeled responders or cells activated in same culture condition as the suppressor cells being tested. Row C contains a positive control consisting of responders plus a decreasing number of freshly sorted Tregs ($CD4^+CD127^-CD25^{high}$). Row D contains the suppressor cells being tested. Depending on how many populations are being tested, the last row should be reserved for unlabeled responders only activated with APCs to use for compensations on the flow cytometer if necessary.
3. Using either flat or round bottom 96 well culture plates, add 50 μl complete media to each well (see Note 13).
4. Pipet 50,000 negative control cells (50 μl) into the first three wells (columns 1–3) of Row B. Do the same for the positive control (Row C, columns 1–3) and suppressor cells being tested (Row D, well 1–3).

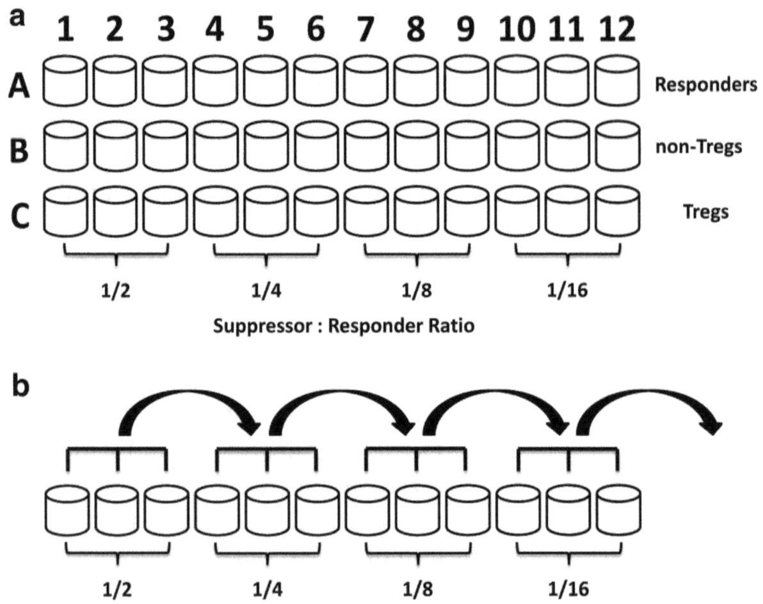

Fig. 1. Schematic representation for setting up the suppression assay with 96 well plate. (**a**) Truncated representation of 96 well plate with 12 columns labeled by numbers and 8 rows labeled by alphabets. Assay is performed in triplicate wells. (**b**) Figure demonstrates how to perform a dilution of the non-Treg and Treg cell number relative to responders.

5. Set a multichannel pipette to 50 μl volume. Mix the cells in the first three wells of each row by pipetting up and down at least 10 times and then transfer 50 μl (~25,000 cells) into the next three wells (4–6) (Fig. 1b). Repeat the mixing and transfer 50 μl (~12,500 cells) into the next three wells (7–9), then repeat for the final triplicates in wells 10–12. Discard 50 μl after completion of mixing. Now each of the rows, except for the first (Row A) and last, should contain 50 μl volume with 25,000 cells in wells 1–3 (1/2 ratio), 12,500 in wells 4–6 (1/4), 6,250 in wells 7–9 (1/8) and 3,125 in wells 10–12 (1/16).
6. Add the CFSE-labeled responders that have been resuspended at a concentration of 1×10^6 cells/ml. Using a repeater pipette, add 50,000 responders (50 μl) into Row A (CFSE-labeled responders only) and the other rows containing the negative, positive and tested cells. Pipet 50,000 unlabeled responders (compensation cells) into the last row. Total volume for all the wells should be at 100 μl.
7. Add the HLA-DR$^+$ APCs containing the appropriate amount of anti-CD3 Abs to give a final concentration of 0.25–0.5 μg/ml per well (see Note 14). With a repeater pipette, add 25,000 APCs (50 μl) to all the wells (see Note 15).
8. Pipet 50–100 μl of complete medium into each well to give a final volume of 200 μl/well.

9. Incubate the covered plate for 3–5 days at 37°C and 5% CO_2. Inspect the wells under the microscope each day to monitor cell interaction.
10. On day 3, remove cells from the first well of the responders only (Row A) and analyze by flow cytometry for the degree of cell proliferation based on CFSE dilution. In general, the assay should stop when over 50% of responders have divided at least 3–5 times.
11. Once the responders have proliferated sufficiently (Row A), transfer cells from triplicate wells from the rest of the plate into 5 ml flow tubes. Stain/wash the cells with anti-CD4 fluorochrome conjugated Abs (see Subheading 3.1) to identify any potential CFSE-labeled responders that have fully diluted their CFSE into the unlabeled population (see Notes 16–18).
12. If a flow cytometer is not available, proliferation and suppression can also be quantified using tritiated thymidine (^{3}H-TdR) incorporation and detection with a scintillation counter. Actively proliferating cells typically take-up and incorporate thymidine as they synthesize DNA. Pulse the wells with 1 μCi (0.037 MBq)/well for the last 6–8 h of the assay, then harvest the plate on to filter paper to measure tritiated thymidine incorporation on the scintillation counter.

3.5. Analyzing Flow Cytometric Data (see Note 19)

1. Start first with the responders only and gate on the viable cells based on a forward and side scatter plot (Fig. 2a).
2. From that gated population, set a plot of side scatter (SSC) versus CFSE or CD4 versus CFSE and gate on the CFSE positive cells (Fig. 2b).
3. Draw a histogram of the CFSE positive cells to quantify the number and percentage of division (Fig. 2c, d). The first peak on the right side of the histogram is typically the undivided cells, signifying zero divisions. The first peak to the left of the undivided peak is considered the first division. The next peak after that is the second division and so forth. Most analysis programs can also analyze the percentage of cells in each peak and the total percentage of all divided cells. Occasionally, the peaks of division are not that discrete and might be difficult to quantify number of division (Fig. 3).
4. After this analysis is completed for the responders, apply the same parameters to analyze the level of suppression in the negative non-Tregs (see Note 20), positive Tregs and sample suppressor cells. Use the total percentage of divided cells to measure the level of suppression. Make sure that the gate for the CFSE positive in the responders does not cross into the unlabeled cells in the coculture, because the unlabeled cells will be included in the analysis and will increase the percentage of divided responder cells.

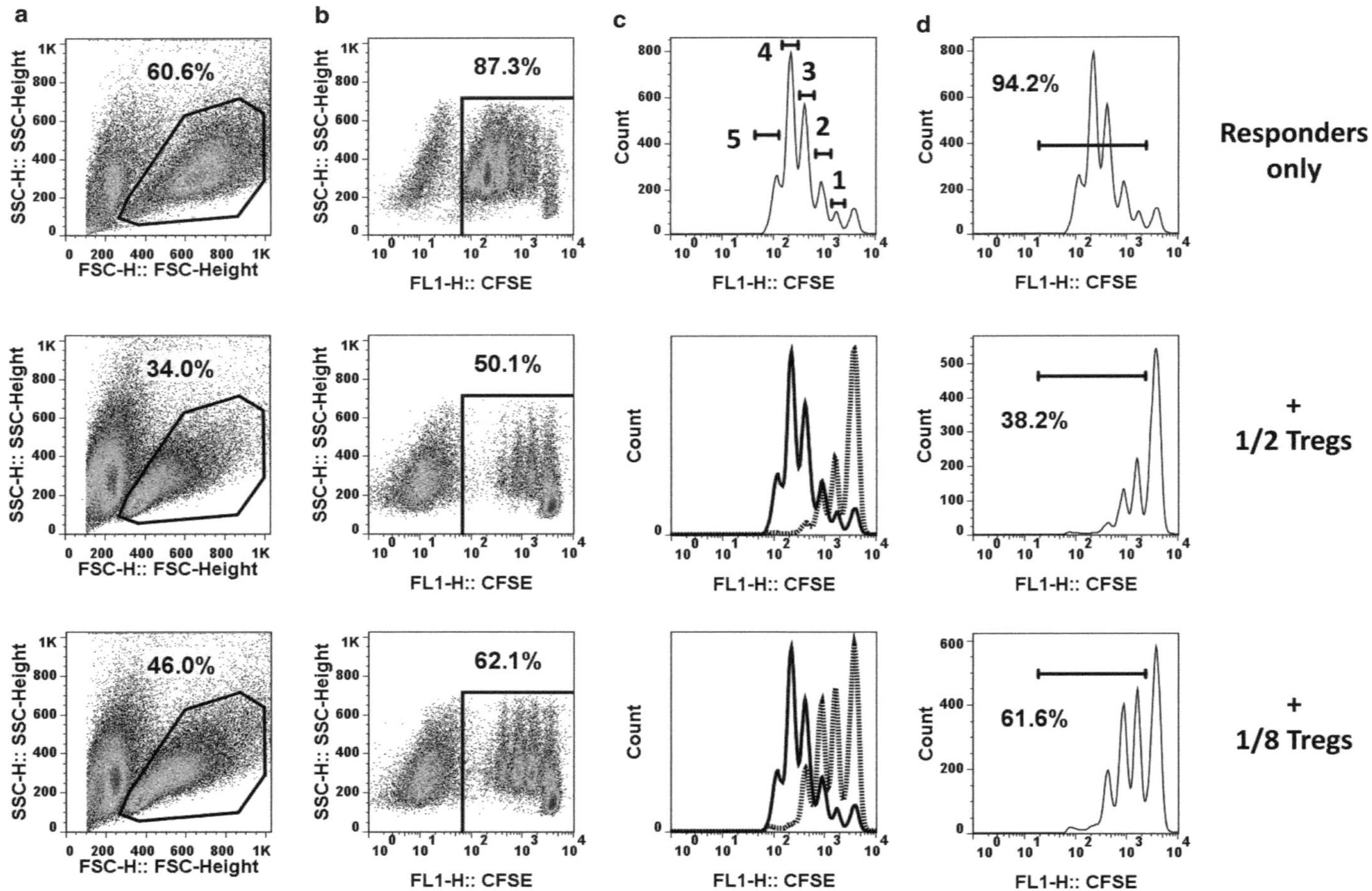

Fig. 2. Flow cytometric plots of CFSE-based proliferation in a 5 day culture. (**a**) Forward (FSC) and side (SSC) scatter plots and gating strategy of the responders only, responders plus 1/2 Tregs and responders plus 1/8 Tregs. (**b**) Gating strategy for CFSE$^+$ cells. (**c**) Histogram of CFSE demonstrating the numbers of division within the responders only (*solid line*) and the overlay plots with the level of suppression (*dashed line*) by the 1/2 and 1/8 Tregs. (**d**) Histogram showing the percentage of CFSE divided cells in the responders only, responders plus 1/2 Tregs and responders plus 1/8 Tregs.

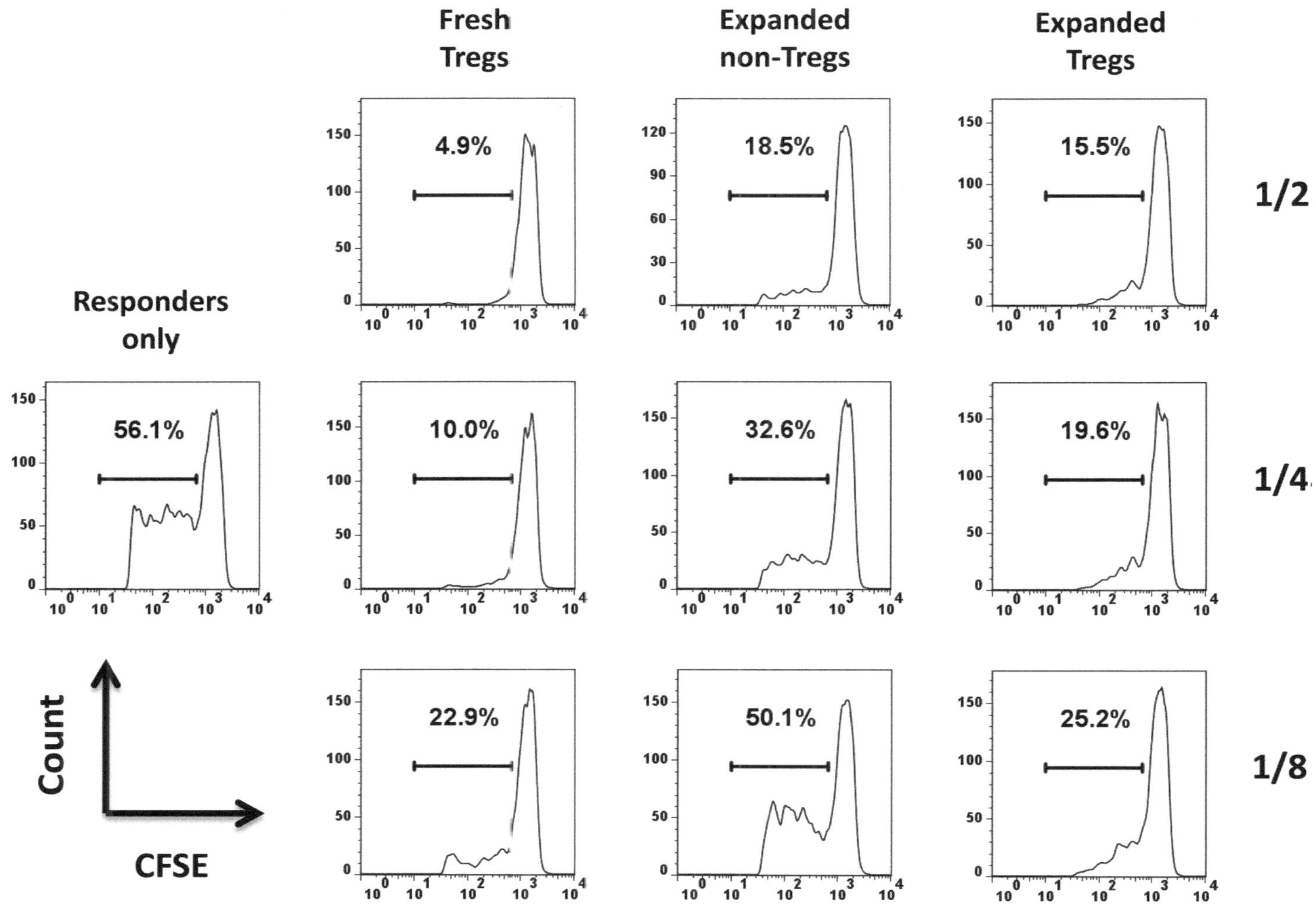

Fig. 3. Flow cytometric plots demonstrating the degree of suppression by fresh Tregs ($CD4^{+}CD127^{-}CD25^{high}$), expanded non-Tregs ($CD4^{+}FOXP3^{-}$) and expanded Tregs ($CD4^{+}FOXP3^{+}$) at varying cell numbers relative to responders only in a 5 day culture.

4. Notes

1. This isolation kit is available from different companies such as Miltenyi Biotec, Stemcell Technologies, R&D Systems and Invitrogen.
2. An automated magnetic cell sorter, such as the autoMACS (Miltenyi Biotec) or RoboSep (Stemcell Technologies), will make separation more convenient particularly with high cell numbers (i.e., apheresis products) and multiple samples.
3. There are untouched CD4 isolation kits available that contained bead-conjugated Abs to all other cell types in the PBMC to provide purified $CD4^+$ T cells that have no bound Abs by negative selection. Alternatively, there is a kit by Stemcell Technologies (RosetteSep Human CD4+ T cell Enrichment Cocktail) that utilizes immunorosette formations by crosslinking unwanted non-CD4 cells in whole blood or buffy coat to red blood cells for depletion during Ficoll separation, leaving behind untouched $CD4^+$ T cells in the buffy layer. This kit saves one step by isolating $CD4^+$ cells during the Ficoll process. In all kits that purify CD4 by negative selection, it is critical to save an aliquot of whole blood or buffy coat for isolation of HLA-DR^+ APCs; otherwise they will be depleted during the CD4 isolation step.
4. To obtain the highest yield and purity of Tregs, it is important to use the fluorochrome with the highest fluorescent intensity for CD25, such as phycoerythrin (PE) or Brilliant Violet 421. Do not label the cells with FITC so that CFSE, which has excitation/emission maxima of ~492/517 nm, can be used to stain the responder cells in order to track their proliferation. Likewise, do not use Brilliant Violet 421 if CellTrace Violet stain (Invitrogen) is utilized to monitor cell proliferation. Depending on the source of the antibodies, the amount should be titrated to determine optimal concentration.
5. If a flow cytometric cell sorter is not available, an alternative method for the isolation of responders and Tregs is magnetic-based cell sorting. There are several Treg isolation kits available from different companies. A major issue with this method over flow cytometric cell sorting is the variability in Treg purity. The kit that incorporates CD127 has higher FOXP3 purity for Treg isolation. Reserve an aliquot of 10–25% of the PBMC for isolation of APCs depending on the quantity needed for the assay. Start with a kit that allows for the negative selection of untouched $CD4^+$ T cells by depleting all other cell types. After isolation of CD4 either through magnetic columns or an automated machine, repeat the process with anti-CD127 magnetic beads to purify $CD4^+CD127^+$ responder T cells. In the $CD4^+CD127^-$ fraction,

use anti-CD25 beads to isolate $CD4^{+}CD127^{-}CD25^{high}$ Tregs. The amount of CD25 beads will determine the quantity and purity of $FOXP3^{+}$ Tregs. The higher the quantity of beads, the greater the quantity of $CD25^{+}$ cells but the lower the percentage of FOXP3 due to the isolation of cells with low expression of CD25. In addition, the flow rate and strength of magnetic force through the column will influence the purity and yield of Tregs.

6. Since CD25 is an activation marker that is upregulated upon T cell activation, one factor that is difficult to control is the variability in CD25 expression among donors depending on their inflammatory or infectious status. The level of CD25 signal in the $CD127^{+}$ gate can provide guidance for determining the cutoff of $CD25^{high}$.
7. This step is critical for determining the purity of the Treg population to assist with the interpretation of the level of suppression from the assay.
8. In the murine suppression assay, APCs are obtained by depleting T cells from splenocytes. This APC population contains mostly B cells and few dendritic cells (DCs). In human, CD3-depleted PBMC are typically used as APCs. This population contains a highly variable amount of B cells, natural killer (NK) cells, DCs and monocytes due to the variability among donors. This variability affects the consistency in the activation and proliferation of responders. Since the suppression assay assesses the ability of Tregs to inhibit the proliferation of responders, it is imperative to achieve optimal and consistent T cell proliferation in order to reliably quantify the suppressive function of Tregs. Due to the variability in APCs within CD3-depleted PBMC, a more consistent strategy is the use of HLA-DR^{+} (MHC II) cells as APCs. In general, this population contains around 35% $CD19^{+}$ B cells, 45% $CD14^{+}$ monocytes, 10% $CD16^{+}CD56^{+}$ NK cells, 5% myeloid DCs (Lineage $1^{-}CD1c^{+}$) and 5% plasmacytoid DCs (pDCs, Lineage $1^{-}CD304^{+}$).
9. Miltenyi Biotec sells a kit for positive selection of HLA-DR^{+} cells. To determine the purity after magnetic sorting, 10^{5} cells from the HLA-DR^{+} fraction could be stained with fluorochrome-conjugated anti-HLA-DR Abs in 100 μl FACS buffer for 20 min at 4°C. The cells are then washed with FACS buffer, centrifuged and resuspended in FACS buffer for flow cytometric analysis. Similarly, the cellular composition of the HLA-DR^{+} fraction could be determined by using flourochrome-conjugated Abs for detection of DCs, B cells, monocytes, NK cell and T cells.
10. For visual demonstration, a video has been published demonstrating the labeling process (10). Other cell labeling kits utilizing different dyes are available that can be used to

replace CFSE or to distinguish Tregs from the CFSE-labeled responders. One such kit is the CellTrace Violet Cell Proliferation Kit from Invitrogen which contains an esterase-activated phenolic fluorophore with a succinimidyl ester substituent for linking to cell surface and intracellular molecules possessing an amine group. This dye is excited by a 405 nm violet laser and has an emission peak at ~455 nm. Therefore, it can be used in combination with CFSE to distinguish the Tregs and responders in the coculture. One advantage to this dye over other dyes such as PKH is that it provides equally partitioning between daughter cells during division with similar resolution to CFSE. Labeling with this dye is similar to CFSE and a protocol is provided in the kit.

11. It is important to have effective removal and deactivation of unbound CFSE so that it does not stain the suppressor cells. Moreover, a concentration of CFSE labeling that is too high in the responders may result in high fluorescent emission beyond the flow cytometric detectors and may perturb their function and cause poor activation.
12. A mouse-human in vitro suppression assay has also been developed that allows for manipulation of the human Tregs in the coculture without affecting the responder cells (7). In this system, the responders and APCs are murine and the Tregs are human. Antibodies or reagents specific to human without cross-reactivity to murine could be used to perturb the Tregs without affecting the responders or APCs.
13. Flat bottom is preferred for easier visualization of cell interaction and activation under the microscope during the 3–5 day culture period. Others have used round bottom for lower cell numbers (i.e., 10,000 responders) to facilitate close contact cell interaction.
14. The concentration of anti-CD3 will be determined by the level of stimulation desired. Too strong of stimulation might not yield optimal suppressive sensitivity. However, too weak of stimulation might not yield sufficient CFSE dilution and proliferation of responder cells.
15. Depending on the level of activation, proliferation, and suppression, more or less APCs and anti-CD3 Abs could be used. The key is to not have too strong or too weak of stimulation. The ideal condition is 3–5 divisions of greater than 50% in the CFSE-labeled responders by day 3 so that the assay can be terminated early.
16. Fluorochromes such as fluorescein isothiocyanate (FITC) and Alexa Fluor 488 (AF488) cannot be used because they will overlap with CFSE. Choose allophycocyanin (APC) or Alexa

Fluor 647 (AF647) for anti-CD4 so that no flow cytometric compensation is necessary between CFSE and APC/AF647.

17. If CellTrace Violet stain is used to label the suppressor cells to measure their proliferation, the emission of this fluorchrome is recognized by a different detector on the flow cytometer. There are other dyes available to identify and simultaneously track expansion of the suppressor cells in the coculture. Compensation of the flow cytometer might be necessary depending on the dye combination and emission overlap. If CFSE or CellTrace Violet is the only fluorescent dye in the assay, then the samples can be analyzed on the flow cytometer without preparing compensation tubes. If the coculture contains multiple fluorescent dyes that have overlapping emissions, then compensation tubes (e.g., cells stained with a single fluorochrome or dye) must be prepared for the cytometer.
18. If the suppressor cells have been labeled to distinguish them from the CFSE-labeled responders, intracellular FOXP3 staining can be performed on the coculture to analyze the level of FOXP3 in the suppressor cells after the end of the assay. There are several FOXP3 intracellular staining kits available, but the one from eBioscience is most reliable, using anti-FOXP3 Abs clone 206D, 259D, or 236A/E7. The fixation/permeabilization process will not affect the CFSE staining.
19. A program to analyze the flow data is necessary. There are several programs available including FlowJo (Tree Star, Inc), CellQuest (BD Biosciences), Kaluza (Beckman Coulter), and FCS Express (De Novo Software).
20. It is imperative to establish a negative control such as non-Tregs with titrations of the cells being tested to demonstrate the specificity of Treg-mediated suppression. Avoid 1:1 and even 1:2 suppressor to responder ratio because nonspecific suppression by unlabeled cells could occur due to competitive interaction with the APCs. Even at 1:2 ratios, profound nonspecific suppression by non-Tregs could occur, but it becomes less significant at lower titration of 1:4, 1:8 and 1:16 (Fig. 3). Therefore, it is helpful to include negative control in data presentation to demonstrate specificity of Treg suppressor function.

Acknowledgement

This work was supported by institutional funding from the Department of Pediatrics, The University of Texas Medical School at Houston.

References

1. Sakaguchi S, Miyara M, Costantino CM, Hafler DA (2010) FOXP3+ regulatory T cells in the human immune system. Nat Rev Immunol 10:490–500
2. Rudensky AY (2011) Regulatory T cells and Foxp3. Immunol Rev 241:260–8
3. van der Vliet HJ, Nieuwenhuis EE (2007) IPEX as a result of mutations in FOXP3. Clin Dev Immunol 2007:89017
4. Miyara M, Sakaguchi S (2011) Human FoxP3(+) CD4(+) regulatory T cells: their knowns and unknowns. Immunol Cell Biol 89:346–51
5. Cvetanovich GL, Hafler DA (2010) Human regulatory T cells in autoimmune diseases. Curr Opin Immunol 22:753–60
6. Sakaguchi S, Wing K, Onishi Y, Prieto-Martin P, Yamaguchi T (2009) Regulatory T cells: how do they suppress immune responses? Int Immunol 21:1105–11
7. Tran DQ, Glass DD, Uzel G, Darnell DA, Spalding C, Holland SM et al (2009) Analysis of adhesion molecules, target cells, and role of IL-2 in human FOXP3+ regulatory T cell suppressor function. J Immunol 182:2929–38
8. Tang Q, Adams JY, Tooley AJ, Bi M, Fife BT, Serra P et al (2006) Visualizing regulatory T cell control of autoimmune responses in non-obese diabetic mice. Nat Immunol 7:83–92
9. Onishi Y, Fehervari Z, Yamaguchi T, Sakaguchi S (2008) Foxp3+ natural regulatory T cells preferentially form aggregates on dendritic cells in vitro and actively inhibit their maturation. Proc Natl Acad Sci U S A 105: 10113–10118
10. Quah BJC, Parish CR (2010) The use of carboxyfluorescein diacetate succinimidyl ester (CFSE) to monitor lymphocyte proliferation. J Vis Exp 44:e2259

INDEX

Andrew L. Snow and Michael J. Lenardo (eds.), *Immune Homeostasis: Methods and Protocols*, Methods in Molecular Biology, vol. 979, DOI 10.1007/978-1-62703-290-2,

T

V

W

X

MIX
Papier aus verantwortungsvollen Quellen
Paper from responsible sources
FSC® C105338

If you have any concerns about our products,
you can contact us on
ProductSafety@springernature.com

In case Publisher is established outside the EU,
the EU authorized representative is:
Springer Nature Customer Service Center GmbH
Europaplatz 3, 69115 Heidelberg, Germany

Printed by Libri Plureos GmbH
in Hamburg, Germany